MW00562979

BITE

ALSO BY BILL SCHUTT

Dark Banquet: Blood and the Curious Lives of Blood-Feeding Creatures

Cannibalism: A Perfectly Natural History

Pump: A Natural History of the Heart

ALSO BY BILL SCHUTT AND J. R. FINCH

Hell's Gate

The Himalayan Codex

The Darwin Strain

BITE

An Incisive History of Teeth, from Hagfish to Humans

BILL SCHUTT

illustrations by

PATRICIA J. WYNNE

ALGONQUIN BOOKS OF CHAPEL HILL 2024

Published by
Algonquin Books of Chapel Hill
Post Office Box 2225
Chapel Hill, North Carolina 27515-2225

an imprint of Workman Publishing
a division of Hachette Book Group, Inc.
1290 Avenue of the Americas
New York, NY 10104

Printed in the United States of America.
Design by Steve Godwin.

Library of Congress Cataloging-in-Publication Data
Names: Schutt, Bill, author. | Wynne, Patricia, illustrator.
Title: Bite : an incisive history of teeth, from hagfish to humans / Bill Schutt ; illustrations by Patricia J. Wynne.
Description: First edition. | Chapel Hill, North Carolina : Algonquin Books of Chapel Hill, 2024. | Includes bibliographical references and index. |
Identifiers: LCCN 2024005154 | ISBN 9781643751788 (hardcover) | ISBN 9781643756158 (ebook)
Subjects: LCSH: Teeth—Evolution—Popular works. | Teeth—Popular works. | Dentition—Popular works. | Adaptation (Biology)—Popular works.
Classification: LCC QL858 .S38 2024 | DDC 599.9/43—dc23/eng/20240405
LC record available at https://lccn.loc.gov/2024005154

10 9 8 7 6 5 4 3 2 1
First Edition

With deep respect, admiration, and gratitude,
this book is dedicated to my academic and scientific mentors:
John Hermanson (Cornell University),
Nancy Simmons (American Museum of Natural History),
and Edwin Spicka (SUNY Geneseo).

TEETH! They are very much in style
They must be very much worthwhile!
—Dr. Seuss

Contents

BITE

Introduction

When humans are confronted by large numbers of seemingly disorganized items, they will frequently do two things—that is, if they don't turn and run the other way. First, they'll separate the stuff into groups based on similar appearance. Next, if those groups haven't been named already, they'll name them. Basically, then, in much the same way that an anonymous launderer from the past used color or the lack of it to divide a large pile of dirty clothes into two smaller piles—lights and darks—the eighteenth-century Swedish physician Carl Linnaeus (1707–1778) divvied up a dizzying menagerie of animals into vertebrates and invertebrates. Here, though, the sorting depended on whether or not the critters were sporting a vertebral column, commonly referred to as a backbone.

For his trouble, Linnaeus was christened the Father of Taxonomy, the science of separating and categorizing living things. The identity of the Champion of Darker Darks and Lighter Lights remains lost to history.

Now that we've got that down, let's say you were asked to list the unique features found in the vertebrates—a group (or taxon) composed of fish, amphibians, reptiles, birds, and mammals. The vertebral column that gave them their name would be a strong first choice. Also aptly known as the spinal column, it surrounds and protects the delicate spinal cord, while providing a sturdy attachment site for muscles, ribs, limbs, and the head. Because the vertebral column is not a solid rod but is instead formed by a varying number of chain-link vertebrae, it is also flexible enough to play a key role in movement and locomotion.

Another strong candidate for our list of uniquely vertebrate features would be bone, a form of connective tissue made up of varying amounts of stony minerals (mostly calcium-rich hydroxyapatite), bendy rods of the ubiquitous protein collagen, and water. Bone tissue forms an array of skeletal structures, conveniently—though at times, confusingly—known simply as bones. A caveat here is that while most vertebrates have bony skeletons, sharks and their flattened relatives the skates and rays have done quite nicely for the past 450 million years or so with skeletons (including vertebral columns) composed of cartilage—which is essentially bone without the stony mineral salts. And though sharks apparently never *had* bone, there are also a few vertebrates, like the hagfish and lampreys, that lost theirs somewhere along their respective evolutionary paths to corpse munching and parasitism. For these reasons, bone occupies a spot slightly below the vertebral column on my personal list of distinctly vertebrate traits.

"What about brains?" you might ask, utilizing some of the eighty-six billion neurons that make up the human model. But after checking with official scorers back at the studios in New York, that particular feature gets the thumbs-down. The reason is that unlike the vertebral column and bone, possession of a brain isn't limited to vertebrates. These complex organic computers evolved much earlier in invertebrate groups like insects, crustaceans, and spiders. Eventually, brains were passed on from ancient invertebrates to the ancestors of the first creatures with a vertebral column. Finally, although possession of a large brain is often considered a vertebrate trait, it should be noted that the brains of both cats (vertebrates) and octopuses (invertebrates) contain roughly two hundred million neurons.*

Like the vertebral column and bone, there is yet another vertebrate

* Dog brains contain around twice as many neurons.

innovation that sits near the top of our list: teeth. Teeth are *that* important, I will argue, because much of the diversity and long-term evolutionary success of the vertebrates can be attributed to their presence. The appearance of teeth, around five hundred million years ago, and the serious remodeling that occurred after that, enabled myriad forms of vertebrates to obtain and process food in pretty much every conceivable environment—from sun-torched deserts to oceans and rain forests teeming with thousands of species of animals and plants. This matching of organism and environment forms what is otherwise known as an ecological niche. And the ability of organisms, including plants, to successfully fit into and exploit a particular niche depends on their having evolved a suite of characteristics called adaptations. Most adaptations involve some aspect of anatomy (our molars, for example, with their broad, flat surfaces) or the behavior related to those anatomical features (like using our molars to crush or grind solid or tough food items into softer, more easily digestible bits).*

But tooth functions extend far beyond procuring and processing food. These hard, mineral-rich structures, which include variations such as tusks and fangs, often play a crucial role as defensive weapons. According to the World Health Organization, over one hundred thousand people die each year from snakebites—along with three times that number who end up with amputations and other disabling conditions. And one needs only the briefest glimpse of the upper canines of snarling dogs or baboons to know that teeth are an efficient means of expressing strength, dominance, and aggression. Less obvious is their involvement in mating behavior and even parent/offspring relationships, as when

* Adaptations give those with them a survival edge over those without them. In doing so, they also provide individuals with a greater probability of surviving long enough to mate and pass on their genes (and their adaptations) to the next generation.

young bats use hook-tipped baby teeth to hang on for dear life as their moms navigate the night skies.

Teeth also enable animals to manipulate and change their environments. Tree-trunk-gnawing beavers and water-hole-excavating elephants may come to mind, but researchers examining tooth wear patterns have shown that Neanderthals and early modern humans also used their teeth as tools, to soften tough fibers and animal hides—a practice that continues in some cultures.

Human teeth also project information about the individual. A mouthful of well-formed teeth can portray vitality, power, wealth, and success, while misshapen or absent teeth can lead to the opposite assumption. By the time George Washington was sworn in as president in 1789, he had only a single tooth—a dismal dental condition that could have done serious damage to his standing as the leader of a newly minted nation. Over the last two decades of his life, Washington would go through at least four sets of bulky, ill-fitting dentures. And though much improved from the painful-to-wear appliances of the eighteenth century, dentures remain a source of consternation to this day.

But false teeth, dentures, and the like have been around far longer than the time of the Founding Fathers. Fashioned by ancient Etruscan artisans nearly three thousand years ago in what is now Italy, these pre–Roman Empire dental appliances were composed of filed-down animal and human teeth, and, as we'll discover, they served a surprising purpose.

Back in the animal kingdom, we'll explore why the term "specialized diet" can be a serious understatement—as in one scale-munching fish species with jaws adapted for plucking scales from either the right side or left side of a prey's body, but not both sides. From an evolutionary standpoint, one key benefit for the specialist is the ability to eat things that potential competitors aren't able to eat. Vampire bats stand as a prime example, with each of the three genera using razor-sharp teeth and an assortment of unique adaptations to feed on a widespread but generally untappable resource: blood.

In contrast to those vertebrates with specialized diets are the omnivores—creatures able to consume a wide variety of food sources, though the term generally implies a diet consisting of plants and animals.

But whether a tooth belongs to a specialist or an omnivore, most of its bulk (including the jaw-gripping roots) is made up of dentin, a very bone-like noncellular mixture of hydroxyapatite, tiny tubes of collagen, and water. The tooth's crown (the part that sticks out from the gums) is covered by a thin layer of enamel, the hardest substance found in the vertebrate body. Additionally, all teeth contain a central chamber known as a pulp cavity, which contains blood vessels and nerve endings.

Beyond the importance of teeth to their owners, much of what we know about ancient life on this planet has come from the study of teeth. While other body parts decompose, teeth are far more likely to endure the test of time, in some cases for hundreds of millions of years, as fossils. Once the tooth of a deceased animal is buried (usually under an accumulation of sediment), it is protected from the effects of general wear and tear, weathering, and erosion. Over time, water, seeping down through the soil, enters the tooth through microscopic pores in the enamel. The water carries with it minerals like silica (think sand, concrete, and glass) and calcite (think limestone and marble). In turn, these substances are transported deeper into the tooth through channels in the dentin. Eventually, the minerals solidify, leaving the tooth even harder than it was in life, and making teeth the most common vertebrate fossils by a wide margin.

Beginning in the nineteenth century, fossilized teeth enabled paleoanthropologists to begin unraveling the mysteries of human evolution. More recently, they've provided information on ancient diets, health, climates, and ecology. The tales told by teeth currently stretch from early humans to ancient civilizations and beyond, through periods of famine, war, disease, and starvation. Today, researchers are exploring how tooth-related techniques used to examine life events in the distant past might also be used to identify early-life stress exposure that can put individuals at risk for mental health problems.

Though extremely durable, teeth are not indestructible, as those who discovered the pleasures of refined sugar would learn beginning in the seventeenth and eighteenth centuries. Effective preventive care was then unknown, as was the concept of sterile conditions, and dental quackery was widespread and dangerous. An entry found in a London bill of mortality for a single week beginning on August 15, 1665, listed 111 people who had succumbed from tooth-related issues.

Thankfully, things have gotten better. That said, many people were shocked in early 2023 by nightmarish photos depicting the aftereffects of a widely used but unregulated dental device called an anterior growth guidance appliance (or AGGA). Attracted by claims that it could correct bite issues or overcrowded teeth and eliminate sleep apnea, all without surgery, thousands of hopeful patients shelled out an average of $7,000 to have one of these wire contraptions installed. Unfortunately, instead of health benefits, many patients received damaged gums and jaws, exposed tooth roots, and flared or lost teeth. Accordingly, the use of AGGAs has fallen off dramatically, since most dentists are no longer promoting them.

Controversial dental appliances aside, given the importance of dental health (now a multibillion-dollar industry), improvements in both preventive and restorative care continue today, with promising new therapies being developed around the world. In one such avenue of research, scientists played the role of rodent dentists. But instead of placing fillings into cavity-ridden rat teeth, they inserted stem cells into the holes they'd drilled. The results were spectacular, namely the production of new dentin by the stem cells. In another study, researchers are investigating the potential to stimulate the growth of a second set of adult teeth. The hope is that this and related advances in dental medicine will soon make typical tooth fillings obsolete and dentures a thing of the past.

Teeth of varying sizes, shapes, and functions have been integral in allowing vertebrates to take full advantage of the wealth of resources found on our planet. Not to be forgotten, though, are the microscopic organisms that live on, around, and below our teeth. Divvying up the resources (a.k.a. resource partitioning) also exists in these microenvironments, as does competition and even cooperation between different microscopic organisms. Today, investigators are looking into a link between the presence of a common bacterium (*Porphyromonas gingivalis*) found in infected gums and the development of Alzheimer's disease.

Before we get to that, though, and before addressing some seriously interesting questions about the origin of vertebrate teeth, we'll put to rest any notion readers might have that "a tooth is just a tooth"—though hammering that notion into the ground like a tent peg might be a better descriptor.

PART I

Toothy Adaptations in Nature: The Specialists

[1]

Vampire Bats Don't Suck

These bats are neither more nor less than those here, and are accustomed to bite at night, and commonly bite mostly on the end of the nose or fleshy part of the fingers or toes. They take so much blood that it is something that can't be believed without seeing . . .

—Gonzalo Fernando de Oviedo,
sixteenth-century soldier of fortune

One of my favorite things about social media is connecting with people I haven't seen or heard from in decades. When some of these old acquaintances contact me and we get around to career talk, I tell them that I've been teaching anatomy (though I recently retired), working as a research associate at the American Museum of Natural History, and studying bats for over thirty years. With my long-standing love of animals and the macabre, no one seems surprised that with roughly 1,470 species of bats to choose from, I picked the three species of vampire bats as my research subjects. And with their unique feeding behavior (they're the only vertebrates that feed solely on blood) and just-as-unique dental anatomy, these fascinating and long-misunderstood members of the mammalian order Chiroptera have had a lot to tell us about teeth.

In the spring of 1991, and not long after entering my doctorate program in zoology, I started thinking about potential topics for thesis-related research. As these sorts of things tend to go, I was expected to pick a topic that fell under the expertise of the graduate committee chairperson. In my case, it was John Hermanson, a young faculty member whose only other grad student had departed the year before. Hermanson, who taught anatomy at Cornell University's veterinary school (though he was not himself a vet) had a strong background in muscle biochemistry, especially variants of the famous contractile protein duo actin and myosin. In the presence of calcium and adenosine triphosphate (the energy currency of the cell, better known as ATP), the chain-like molecules of actin and myosin ratchet past each other. Occurring simultaneously millions of times over, this molecular movement produces the muscle contraction responsible for actions like walking and galloping—activities formerly popular with the horses whose leg muscles Hermanson kept stored in an ultracold freezer.

Hermanson was at the forefront of the relatively new field of functional morphology, a modern take on classical anatomy. Practitioners of "func morph" studied the relationship between the structures of an organism (i.e., its anatomy) and the function of those structures.* Hitting the literature hard, I quickly became intrigued. It also became apparent that Hermanson's favorite study animals were clearly off the beaten horse track—some of them in fact flitting over said tracks after dusk.

Having spent at least a third of my childhood free time peering under rocks and logs for snakes, salamanders, and other creepy-crawlies, another third maintaining a collection of exotic pets (I had a monkey named Guggie and, later, a boa constrictor named Alice), and the remainder of my time watching horror movies, it took me approximately five seconds to choose bats over biochemistry.

---

* Although "anatomy" (from the ancient Greek "to cut up" or "to cut open") and "morphology" (ancient Greek for "the study of form") are certainly related, anatomy deals with the presence and the description of structures and their parts in an organism. Morphology is concerned with the size, shape, and structure of organisms and their parts, as well as relationships between those parts in a particular organism or between those parts in different organisms.

WHEN THE FIFTEENTH- and sixteenth-century Europeans returned from their flag-planting exploits in the New World, along with the loot they had pillaged and the Indigenous people they had kidnapped and enslaved, they also brought back tales of strange creatures—of sea monsters, of cyclopean humans with tails, and of bats that attacked men in the night and drank their blood. The latter were generally described as hideous creatures sporting five-foot wings. With these Westerners far more concerned with gold, God, and geography, it would take another 250 years for the first naturalists to begin closely examining and classifying the natural wonders they either encountered in places like Central and South America or pulled out of bottles and shipping crates returned from those regions.

When it came to unmasking the identity of the vampire bats, though . . . they blew it. Undoubtably, most of these scientist types saw only the results of vampire bat attacks, which took place in the dead of night, but not the attacks themselves. And the majority of the accounts they collected were likely secondhand tales or worse. Many of the bats they did see had spear-shaped nasal structures projecting vertically from their snouts. Unfortunately, this led the naturalists to the mistaken belief that these nose leaves were wielded like fleshy stilettos. The thought was that after impaling their victims, the vampires would then drain their blood through the resulting gash.

Taxonomists began assigning scary-sounding names to any specimens with a nose leaf, names like *Vampyrops*, *Vampyressa*, *Vampyrodes*, and *Vampyrum*, though, in truth, none of these bats were vampires. Along the way, the group picked up the common name New World leaf-nosed bats. Things went further off the rails as the presence of nose leaves in two Old World bat families (Rhinolophidae and Megadermatidae) likely contributed to the incorrect belief that vampire bats could also be found throughout Europe, Africa, the Indo-Pacific, and Asia.

Today, vampire bats belong to the large (with approximately 170 species) and diverse family Phyllostomidae (in which all but three species do *not* feed on blood). Though they are still commonly referred to as New World leaf-nosed bats, we now understand that nose leaves aren't weapons but are instead employed, megaphone-style, to direct the larynx-generated echolocation calls that characterize the bats that possess them.* I've always found the earlier interpretation of nose-leaf function more than a bit odd, basically because nose leaves are soft and pliable, and in no way capable of inflicting wounds of any kind. Perhaps it was the permanently stiffened condition they were in when they arrived at museums and universities across Europe that misled taxonomists.

It is nearly always a surprise to non–bat biologists that there are but three species of blood-feeding bats alive today. Their ranges are confined to parts of Mexico, Central and South America, and two Caribbean islands (Trinidad and Margarita). The fact that they lack elongate nose leaves, which aren't a modern-day requirement for inclusion in the Phyllostomidae family, but instead possess pad-shaped versions of these structures, goes a long way to explain the nonvampiric scientific names the trio was assigned in the nineteenth century: *Desmodus rotundus*, *Diphylla ecaudata*, and *Diaemus youngi* (see figure 1). But even without sinister-sounding scientific names, each of these bats is wonderfully equipped for a life of obligate sanguivory (blood feeding only)—and several of their adaptations are related to their choppers.

Vampire bats have the fewest teeth of any bat species:† twenty-six in *Diphylla ecaudata*, the hairy-legged vampire bat; twenty-two in *Diaemus youngi*, the white-winged vampire bat; and twenty in *Desmodus rotundus*, the common vampire bat. *Desmodus* is also the only bat with a

---

* The order Chiroptera is divided into two suborders, the Yangochiroptera (most echolocating bats), made up of approximately 1,041 species, and the Yinpterochiroptera (which includes the "flying foxes"), with 482 species.

† The greatest number of teeth found in a bat is thirty-eight.

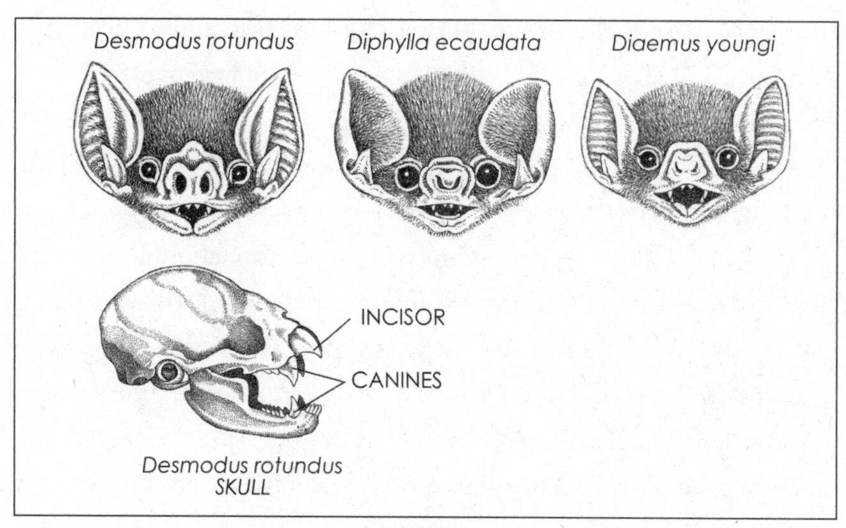

FIGURE 1

single upper and lower molar on each side of the jaw. Interestingly, the fact that vampire bat molars are few in number and tiny in size makes perfect sense. Molars generally serve as the functional equivalent of grinding mills. In many mammals, their job is to smash food items into near pulp—which not only increases the surface area available for digestive enzymes to start doing their jobs but makes it easier to swallow that saliva-moistened mouthful of whatever. Vampire bats consume no solid food, so molars are not required—especially given the metabolic expense of growing and maintaining them.

Premolars (located in back of the incisors and canines but in front of molars) usually see action before molars, shearing large bits of food into smaller ready-to-be-mashed bits. But in vampire bats, they act like a barber's razor, clearing away tiny patches of their prey's hair, feathers, or scales, in preparation for a bite.

Not long after I began studying these creatures, I started to wonder who their ancestors might have been and what they were doing before they made a switch to full-time blood feeding. Although vampire bats clearly had their origins in what is now South America, there was little in the way of a fossil record indicating how they might have evolved.* With no fossils to point the way, a couple of hypotheses had been proposed, each suggesting how vampire bat ancestors (protovampires) might have been feeding.

Zoologist Dennis Turner addressed the question briefly in 1975. "Perhaps a progenitor of today's vampire bats specialized on the ecto-parasites of larger wild animals," he wrote, likely referring to blood-filled ticks as a food source before a transition to blood obtained directly from animals that the ticks were feeding on.

In the wound-feeding hypothesis proposed by renowned bat biologist Brock Fenton, protovampires likely fed at previously existing wound sites commonly found on large animals. There, the stealthy fliers munched on the maggots and other insect larvae that called those festering gashes home. Importantly, though, they would have also gotten a taste of blood from the injured animals—eventually making the switch from feeding on wounds to feeding on blood obtained from bites they themselves had inflicted.

Primarily because neither of these feeding styles is exhibited by any known bat species, I came up with an alternative hypothesis on vampire bat origins. This one was based on a form of feeding behavior that *is* present in extant (i.e., currently living) bats, including species that are closely related to the three vampires. The arboreal-feeding hypothesis posits that vampire bat ancestors living in what is now South America were carnivores dedicated either full- or part-time to feeding on victims they attacked in the trees. Back in the ancient forests, roughly ten million years ago, protovampires would have conducted "sneak up and pounce"

---

* Fossils from tropical environments are relatively rare, since bodies generally don't last long enough to be buried by sediment in hot, humid conditions. Combining this with the reality of tiny, delicate bat bones, the odds of fossilization and discovery become Lotto-like.

attacks upon smaller, tree-dwelling prey (like large insects and frogs), but they later tweaked their attacks to allow them to feed upon much larger arboreal species like marsupials, primates, and tree sloths, which also happened to be evolving at the same time and place. Since these animals were too large for the bats to overpower using previous attack strategies, this new predation technique could be described as "sneak up and bite."

According to the arboreal-feeding hypothesis, some ancient vampire bats, having evolved a strictly blood-feeding diet, would have moved down from the trees (stealthily, of course) to take advantage of terrestrial prey, which may have included megamammals like giant sloths and heavily armored armadillo relatives known as glyptodonts. These four-footed blood banks had been around for roughly twenty million years before the first vampire bats are thought to have evolved. Perhaps reflecting the presence of larger prey, several extinct vampire bat species, including *Desmodus stocki* and the wonderfully named *Desmodus draculae*, were significantly larger than the three extant vampires, with *D. stocki* inhabiting what is now Florida.*

In each of these three hypotheses, natural selection would have acted to transform protovampires into the ultimate stealth hunters. Their arsenal included dental modifications that made their bites painless, thus preventing their prey from fleeing or biting back. Other adaptations functioned to keep a victim's blood flowing for as long as possible before clotting. Changes to teeth centered on the evolution of the ultrasharp incisors and canines that characterize living vampire bats. The anticoagulants currently found in vampire bat saliva, which keep the prey's blood

---

* In my 2016 novel, Hell's Gate, I took some creative license with the size of *D. draculae* (which likely survived into modern times), placing the last roost of these amazing creatures in a cave in Brazil's Central Plateau during World War II. There, the raccoon-sized predators (yes, that part was a stretch) menaced Nazis and Allies alike.

flowing, may have evolved from previously existing substances used by the bats themselves to prevent their own accidental blood-clot formation. As we'll soon see, this form of chemical weaponization would mimic the evolution of other naturally occurring substances, like those involved in lowering blood pressure. In many species of snakes, fish, and even a few mammals, these compounds and others were modified through natural selection into venom. Defined as "a toxic substance delivered by a bite, sting, or barb," venom would provide a survival edge to the creatures possessing it—over those that didn't.*

Sadly, there are no transitional fossils showing a clear link between non-blood-feeding leaf-nosed bats and their vampiric descendants. But whether vampire bat ancestors were wound nibblers, tick gleaners, or prey-switching carnivores, one certainty is that their three living descendants are smaller than most people think. A whole lot smaller, in fact. With a body length averaging about 3.5 inches and a wingspan of 7 to 8 inches, vampire bats generally weigh in at two ounces or a shade less—before settling in for a blood meal, that is. The premeal weigh-in is important, since bats in each of these species can consume up to 50 percent of their body weight in blood each night.

To reduce competition where the species overlap (in places like Trinidad and Brazil, where I studied them), vampire bats divvy up the resources, with the common vampire bats feeding primarily on large terrestrial imports like cattle, horses, and pigs, and the white-winged

---

* Quick review: So, what's natural selection again? Basically, it's like artificial selection, in which a pigeon breeder (for example) only allows pigeons with certain desirable traits or characteristics to mate and pass the traits on to the next generation. In natural selection, the environment (not the bird guy) selects the traits that work best in that environment. The individuals possessing these traits are more likely to survive and pass them on to the next generation than the individuals lacking them. But—and this is huge—if the environment changes, different traits might be selected, in much the way that a pigeon breeder could suddenly decide to select a different color as being desirable. Natural selection, therefore, is not evolution (as some folks might think). Instead, it is the engine that drives evolutionary change.

and hairy-legged vampires preferring to dine on bird blood. Given the human-generated flood of livestock into South America, populations of the common vampire bat exploded during the early twentieth century. This led my friend and vampire-bat maven Arthur M. Greenhall to christen them "the pinnacle of specialization and adaptation in the bat world." So successful was *Desmodus* that it seems likely that the white-winged and hairy-legged vampires would have been driven to extinction had their diets not consisted almost entirely of avian blood.

"Blood feeding is a tough way to make a living," I mentioned to *New York Times* science writer extraordinaire Natalie Angier.

"But it's a great job if you can get it," she replied, without hesitation.

Angier was right, of course, since there are all sorts of vertebrates walking, slithering, and flying about, each of them carrying a hot meal for creatures with the right suite of adaptations to exploit the red stuff. But the career of an obligate sanguivore is not without its drawbacks. The problems start with the fact that blood is approximately 90 percent water, and what remains is mostly protein. With their diet supplying little to no fat or carbohydrates, an inability to store energy is a serious disadvantage for vampire bats. As a result, they can starve to death if they don't get a blood meal within forty-eight hours. This is reminiscent of the mouse-like shrews we'll be meeting soon, which can starve to death within four hours, though for very different reasons.

I've always been fascinated with the list of adaptations shared by blood feeders of every ilk—from ticks and leeches to bedbugs and vampire bats. These blood-feeding adaptations include (1) small size: as vampire bats with larger bodies would need to obtain more blood to sustain themselves; (2) stealth: since getting close to and away from a meal without being detected, and possibly killed or wounded, is of considerable importance; (3) a set of ultrasharp teeth or pointy toothlike mouthparts to inflict painless wounds; and (4) a variety of anticoagulants applied during or after a bite to keep the blood from clotting before a suitable quantity of it has been drained and consumed.

Among the dentition-related adaptations, vampire bat teeth—namely, their incisors and canines—are truly spectacular. Though there is some dental variation between the three genera, in each case the upper incisors and canines are oversized and razor-sharp (see figure 1). They also give the appearance of being fused into a single tooth, a feature that gave *Desmodus* its scientific name (since *desmos* is Greek for "bundled" or "fused" and *odus* is derived from the Greek *odous*, meaning "tooth"). Only the lower canines are similarly pointy and sharp, with the lower incisors adapted for grasping the skin or fur of a potential bite victim.

But vampire bat attacks are just as interesting as the bites themselves. The initial approach of the common vampire bat to its unsuspecting prey utilizes a combination of echolocation and agile, un-bat-like terrestrial locomotion, which consists of a variable combination of walking, running, hopping, and jumping. Once the bat gets within a few inches of its prey, it employs specialized heat sensors (found in pits located close to the nose) to detect the warm blood flowing through shallowly situated arteries and veins. Next, the potential wound site is licked in preparation for a bite, during which the action of the incisors and canines removes a small divot of flesh from the unsuspecting victim. Roughly one-eighth of an inch in diameter, the single crater-shaped wound is quite unlike the double puncture marks inflicted by fictional vampires.

Since the 1970s, the literature on vampire bats often mentions that their incisors and canine teeth either lack or lose the enamel coating typically seen in other vertebrates (including other bats). The explanations relate this condition to the vampires' ability to maintain their teeth's scalpel-like cutting edges through a process known as thegosis (tooth-to-tooth sharpening), during which the edges of the upper and lower teeth cross each other and sharpen, each time the mouth closes. Presumably, any enamel present would be lost during this process.

Recently, however, some researchers have questioned the absence of enamel in vampire bat teeth, citing conflicting data as well as observations related to the known feeding behavior of these creatures. According to biologist John Hermanson, "Dentin is a 'softer' material than enamel and could wear down fast. Given that vampires are piercing a tough hide on prey such as cattle, it is hard to imagine that normal dentin would withstand this level of wear. Is [vampire bat] dentin a different or harder composite than in other mammalian teeth, or is there a slight layer of enamel present to sustain the sharp edge?"

These questions suggest that additional studies should be undertaken (Attention, grad students!) to better determine the composition of vampire bat teeth—including premolars and molars, whose makeup has gone unreported.

Added to a file I call "Real Vampires Don't Do That" is the fact that vampire bats do not suck blood after inflicting a bite—they lap it. This is accomplished through the pistonlike movement of the vampire bat's tongue, which shoots out through a midline gap located between the lower incisors and a deep groove on the lower lip. As the tongue moves in and out of the mouth, saliva containing a potent cocktail of anticoagulants is applied to the wound. This effectively keeps the blood flowing into the bat's mouth, along a lengthwise groove located on the underside of the tongue. The fluid movement occurs in much that same way that blood streams into a small-diameter capillary tube you might see in a doctor's office or hematology lab. In both instances, the adhesive forces attracting the blood to the inner walls (of the capillary tube or the bat's narrow tongue groove) are greater than other forces, like gravity. In vampire bats, the result is that the blood is drawn along the tongue's length.

It streams into the mouth via so-called capillary action as the tongue is rapidly applied to and then withdrawn from the wound.

Once the blood is swallowed, the vampire bat's supercharged digestive and excretory systems kick into gear, rapidly separating the water in the consumed blood (with its substantial weight) from the mostly protein nutrients contained in it. So efficient are the bat's kidneys that even as the animal feeds, copious amounts of dilute urine are produced and eliminated. One reason for this amped-up excretory system is to eliminate ammonia, a toxic byproduct of protein digestion. Vampire bats also need to take flight after a meal, often from the ground and at a moment's notice. And like any form of aircraft, weight is a major factor at takeoff. Because of this, vampire bats must eliminate as much water weight as quickly as possible—and they do so by peeing it away.* Leaving the site of their attack awash in crimson-tinted urine and the blood of their prey (which continues to flow due to the anticlotting ability of the bat's saliva), the common vampire bat initiates flight in a spectacular fashion—catapulting itself into the air with the equivalent of a powerful push-up.

Preying primarily on birds, the white-winged and hairy-legged vampires have no need of flight-initiating jumps. Instead, they often feed beneath a perching bird, having flown in silently. After nipping one of the victim's scale-covered toes, the bat consumes its fill of blood, which flows continuously due to a combination of anticoagulants and gravity. Once satiated, the bat simply releases its hold on the bird, or the branch supporting it, and flies off. As with the terrestrially feeding common

* Remarkably, vampire bat kidneys shift into what could be considered a lower gear after their initial production of copious urine. Once a bat has returned to its roost, the kidneys then conserve water (since the bats do not obtain additional water from drinking it). As a result, the urine produced by this process has the consistency and look of tarry, rust-colored paste. It also serves as a clear indicator that vampire bats are roosting in a cave or other structure.

vampire bat, the anticoagulants previously applied to the wound cause the prey to bleed long after the bat has returned to its roost. There the vampire sleeps, grooms, or, in some instances, shares blood with a baby bat. In *Desmodus*, adult roostmates that have failed to secure a meal are also fed by successful hunters. This sharing behavior occurs when blood that has been stored in a tiny outpocketing of the stomach is regurgitated.

A final note about vampire bat bites relates to attacks on humans: they do happen, and they can present the threat of rabies transmission. Although fear of rabies is sometimes overblown (it's estimated that perhaps one in two hundred bats carry the virus) and misinformation is widespread (e.g., it's a myth that rabid bats appear asymptomatic), the lethality of the virus can turn contact with an infected animal into a life-and-death situation. Rabies is nearly 100 percent fatal if not treated with a postbite vaccination. Because of this, a basic rule for everyone, except those recipients of a preexposure rabies vaccination series (which can last for up to three years), should be to avoid all physical contact with any wild or stray mammals. Raccoon lovers, please take note. Adults should also teach children to do the same, as well as not to be afraid to immediately report any animal encounter that they or their friends might have—especially if it involves dogs, bats, raccoons, skunks, or foxes.

For those readers who might be concerned about meeting up with a vampire bat in the wild, I can tell you that those odds are likely on a par with winning a Mega Millions jackpot. There are no wild vampire bats in the United States, Asia, or Europe—and that includes Transylvania. That said, vampire-bat-transmitted rabies still occurs within their New World ranges, though generally where people sleep without screens on their windows and doors, or camp out in the open or in poorly designed shelters. Fortunately, and thanks to the efforts of antirabies crews working in countries like Trinidad, rabies in humans is far more rare than it used to be.*

Of course, specialized diets extend beyond vampire bats. Horses and baleen whales are excellent examples, each having evolved unique features enabling them to consume tough, nutrient-poor grasses or tons of tiny marine crustaceans, respectively. But many lesser-known species are just as interesting. Among the twelve families of fish that practice lepidophagy (scale eating), one species stands out. *Perissodus microlepis* is an African cichlid, popular with tropical aquarium enthusiasts. And while scale eating is clearly a dietary specialization, things get even more extreme when you consider that this species exists in two visually distinct forms, known as morphs. One morph has its jaws and teeth twisted to the right, making it easier to snatch scales from the left flank of its victims. The other morph has its mouthparts bent to the left, in order to . . . Well, you know where this is going. Although considering our next specialist, hidden in the Amazonian mud as it waits to ambush unsuspecting victims, making assumptions might be a bad idea.

* In the early 1990s, information on the behavior of white-winged vampire bats was scarce and, as pointed out to me by members of Trinidad's Anti-Rabies Unit, a lot of it was wrong. For example, although it was reported to be nearly impossible to maintain white-winged vampires in captivity, Trini bat expert Farouk Muradali taught me how to do just that, patiently guiding me through a process that also enabled us to safely transport these critters back to New York. There, they were comfortably housed and studied without casualties (bat and human) for three years before moving on to new homes at a local zoo. A take-home message to young researchers visiting foreign countries: Invite locals to be your colleagues and coauthors. They know their stuff!

[2]

Candirus: Be Careful Where You Go

Very small, but solely preoccupied with doing harm.

—Paul Le Cointe

During the lead-up to and following the release of my earlier book about blood and the creatures that feed on it, I spent months traveling around the country, doing book signings, and giving book-related talks at venues of every ilk. In the PowerPoint presentations I had prepared, I tried to demystify sanguivores—creatures that in addition to their vampiric habits (and partly because of them) often share the label "seriously creepy."

Among those, one species stands above the rest, to the point that I've reserved any discussion of it for mature audiences. There always seem to be a few people in the crowd who've heard of this particular animal before, the presence of these folks detectable by their titters of laughter followed by whispers directed at the uninformed sitting nearby.

On one occasion, I was speaking to a crowd of two hundred or so pest-control specialists attending a convention in Manhattan's Washington Heights neighborhood. This time, as I mentioned the name of the creature, there was complete silence.

Wow, none of these guys has ever heard of the critter, I thought.

Taking full advantage of this rare opportunity, I responded by drawing out the overture a bit longer than usual. In retrospect, perhaps it was more than just a bit. As I went into "mysterious creature mode," I tossed out a few red herrings about how this particular fish and its horrendous feeding habits made it the most feared denizen of the Amazon River basin. Then, once everyone seemed prepared for a segment on what they now figured must be an alternative name for the infamous piranha, I clicked on the next slide.

The response was as instantaneous as it was impressive. Two hundred pest-control experts, many of them hardened by years of working in the most gross-out conditions New York City could offer, let out a collective groan and doubled over in their seats as if performing some bizarre version of the wave.

I remember smiling as I looked out at the bent-over crowd, and thinking, *Now* that *is what I call power.*

In the preface to a nineteenth-century text on Brazilian fish species, German botanist and explorer Karl Friedrich Philipp von Martius wrote:

> I should briefly mention another fish which is dangerous to man. The Brazilians call it *Candiru*; the Spaniards in Maynas [a state in Peru] call it *Canero.* It is impelled by a curious instinct to enter the excretory openings of the human body. Whenever it comes in contact with these openings of persons bathing in the stream, it violently forces its way in, and having entered, it causes constant pain, and even danger of life, by biting the flesh. These fishes are greatly attracted by the odor of urine. For this reason, those who dwell along the Amazon, when about to enter the stream, whose bays abound with this pest, tie a cord tightly around the prepuce and refrain from urinating.

The horror stories gained traction in 1930, when an ichthyologist authored a pair of articles on the candiru for the *American Journal of*

Surgery and a book, *The Candiru*. The written accounts of the candiru, a type of catfish, reportedly invading the urinary tracts of unfortunate humans were widespread and sensational. Preventive advice included instructing those traveling to candiru-infested waters to wear tight bathing suits. They were also warned that if stricken, and if a potion prepared from the unripe fruit of the jagua tree (*Genipa americana*) could not be obtained, the alternative treatment was penile amputation (presumably for those folks who had a penis, that is).

But was there any truth to these traumatic tales? Many of these stories were thirdhand and, as such, not reported by anyone who had actually witnessed the events. In his definitive 2002 book, *Candiru: Life and Legend of the Bloodsucking Catfishes*, ichthyologist and candiru expert Stephen Spotte investigated that very possibility, seeking to determine (among other things) if there existed any solid evidence of their notorious behavior and if so, why candirus might embark on what would ultimately be a one-way trip.

First, though, some background information.

Having met freshwater fish authority Nathan Lujan during his tenure as a postdoc at the American Museum of Natural History, I contacted him soon after he had begun a curatorial position at the Royal Ontario Museum in Toronto. Lujan's research interests had often focused on the freshwater fish of eastern North America, but they extended into tropical South America. I also knew that he was especially knowledgeable about the various species commonly referred to as candirus.

I asked him to give me a rundown on the taxonomy of these notorious creatures—who they were, how many species, etc. At present, there are fifteen genera and about forty-one species of candirus, making up two mostly parasitic subfamilies: Vandelliinae (four genera and twelve species) and Stegophilinae (eleven genera and twenty-nine species). Candirus, in turn (along with six other subfamilies), belong to a larger grouping, the family Trichomycteridae. Commonly known as pencil catfish, due to their narrow, elongate bodies, they are widely distributed through the Neotropics—from Panama throughout South America, as far south as Chile and Argentina.

According to Lujan, candirus are distinguished from even close relatives by having at least nine clusters of sharp, mostly backward-facing odontodes around the head (see figure 2), each cluster located on a different bone or region of bone. For those who might be wondering about the difference between teeth and odontodes, the former are hypermineralized structures generally restricted to the upper and lower jawbones or, in the case of sharks, embedded in their gingiva (or gums). Similar structures found anywhere else on a vertebrate's body are usually (though not exclusively) referred to as odontodes.

Like teeth, odontodes and scales are made primarily of dentin—a hardened yellow-hued mixture of phosphate minerals and collagen. Also on hand in these fishy surface structures is enamel, the superhard mineralized substance that forms the thin surface layer of a tooth above the gumline. In both fish scales and odontodes, the chemical makeup of the enamel is slightly different from that covering the teeth of other vertebrates, like mammals, but similar enough to warrant description as enamel-like or enameloid.

In the candiru, two clusters of spiky-looking odontodes project from each operculum. Also known as gill covers, the opercula protect the feathery gills beneath them. They can often be seen rising and lowering slowly as a fish breathes. Here, though, the opercular odontodes provide secure traction as the candiru invades tight spaces. Another cluster of odontodes is centered just below and behind the mouth.

The reason for the candiru's strange external dentition relates to their equally strange diets. By most accounts, the stegophilines feed on the skin, scales, and mucus of other fish. They are also reported to be necrophages—feeding on carrion. According to Lujan, all of the vandelliines

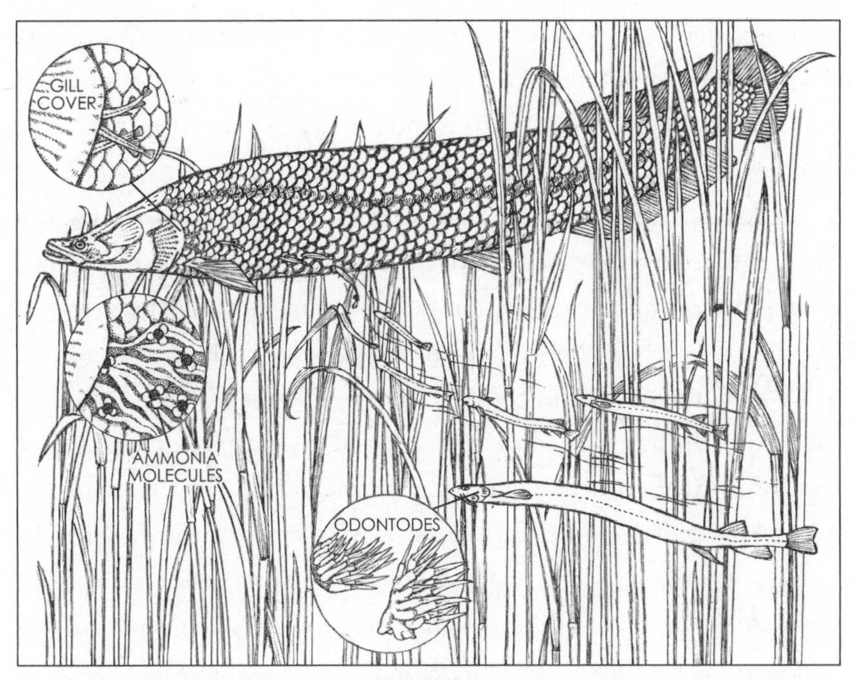

FIGURE 2

are bloodsuckers, though, as it turns out, human urinary tracts are not a preferred dietary destination. When handled, "they are remarkably adept at nosing into the crevices between one's fingers and using their opercular spines to advance through them," he said.

When not being held by researchers, candirus use this ability to weasel their way under the gill covers of the larger fish species upon which they prey. Having entered the gill/opercular cavity, candirus employ a single bite, as well as an arsenal of teeth and variably located odontodes to secure themselves onto the delicate and blood-filled gills of their unsuspecting ride. As a result of the bite, a plug of flesh is removed from the gills, which the candiru swallows. This causes the blood to flow freely from the wound and directly into the candiru's mouth, with the parasite held in place by the backward-facing teeth, odontodes, and its own gill covers, which are expanded outward, umbrella-style. According to Lujan, the blood is actually "ingested somewhat passively and not, strictly speaking, sucked." Of course, this reminded me of the mistaken notion that vampire bats suck the blood of their victims, when, really, the feeding mechanics are more akin to a dog lapping up water.

As for how candirus find their fishy victims, the answer relates to the fact that as vertebrate digestive systems break down protein, they often produce ammonia (NH_3). In most vertebrates of the nonfish variety, the kidneys convert ammonia into urea, which is less toxic. Adding water and other waste substances to the urea further dilutes it, producing urine, which is nontoxic. Urine is stored in an expandable muscular bag, otherwise known as the urinary bladder. Once the volume-driven expansion of the bladder activates stretch receptors located within its walls, a message gets sent to the brain, which is processed into something along the lines of "Find a bathroom." The voluntary response that follows is the elimination of urine through a process known in the trade as micturition, though more popularly known as peeing.

Most fish, including all freshwater species, bypass this energetically expensive process by excreting ammonia directly into the surrounding water, primarily through their gills (see figure 2). Candirus often ply the murkier waters of the Orinoco and Amazon River basins, where eyesight is of little use. Rather than actively searching for their prey, these pint-sized hunters spend much of their time burrowed in the mud—waiting. Like other catfish species, they have an array of chemical sensors located around their head and body that alert them to the presence of specific chemicals in the water. For the candiru, those substances are nitrogenous compounds like ammonia, released by other fish. Improving the efficiency of this sensory system is the widespread placement of chemoreceptors around the body. These provide the candiru with additional information on both the concentration and the direction of the chemical stimulus, allowing the predator to track its potential prey even as it swims—oblivious to the danger. It works something like this: "More nitrogen being detected to the right. Swim toward the right. Now equal amounts of nitrogen being detected on right and left. Swim in a straight line. Concentration of nitrogen extremely high. Prepare for contact with gill cover."

But what about the lurid tales of candirus entering the urethras of unlucky bathers or those who chose the exact wrong place to relieve themselves? According to Stephen Spotte, experiments designed to demonstrate that candirus might be attracted to human urine "yielded completely negative results." The "Fatal Stream Hypothesis" suggests that instead of zeroing in on human urine, candirus might be attracted to currents of water, such as those produced as water exits the gill chamber of a potential prey species. The thought here is that the candiru mistakenly orients itself into a stream of human-generated urine and heads for the source. Then, before you can say "Hey, who turned out the lights?" the pencil-shaped parasite becomes trapped and eventually suffocates, all of this to the horror of the urethra's rightful owner. Once again, this hypothesis lacks anything remotely resembling support. Of note, however, is the fact that any such entry into a human urethra would be the result of an error, since it would be invariably fatal to the candiru.

With nothing in the way of solid evidence for the candiru's notorious and cringe-inducing behavior, these wiener-related tales of woe seemed ready for a slide into the realm of mythology. That is, until October 1997, when a newspaper announced that Anoar Samad, a Brazilian urologist, had removed a candiru from a male victim (referred to as FBC) that he had treated at a hospital in the city of Manaus. According to the newspaper account, Samad said that FBC had been bathing in the Amazon River at Itacoatiara, approximately 110 miles from Manaus, when a candiru "entered in the boy's urethral canal through the penis."

Two years later, when Spotte interviewed Samad, the story had changed, with FBC now claiming that his penis had not been submerged at the time of the attack. Instead, Samad said FBC told him that he had been standing in thigh-high water, urinating, when the candiru "*darted out of the water, up the urine stream, and into his urethra*" [italics not mine]. A short battle reportedly ensued, as FBC unsuccessfully tried to prevent the fish from heading upstream. (I couldn't help imagining the reaction of witnesses to this strange tug-of-war.) Three days later, with his abdomen distended, since he was unable to urinate, FBC underwent surgery to remove the now-deceased catfish. Samad extracted it with an endoscope equipped with tiny clamps known as alligator clips.

As I was writing about the topic a decade later, I ran this somewhat more than slightly modified version of the candiru-leaping-upstream story to my PhD committee member and University of Calgary biomechanics expert John E. A. Bertram. "Could it have happened?" I asked him.

Bertram explained that such a maneuver not only would require the candiru to lift itself out of the water against gravity, but it would have had to then swim faster than the pee stream. The amused researcher told me that this was pretty much an impossibility given the effectiveness of the penis tip as a nozzle capable of generating high-velocity flow. Furthermore, if the candiru *had* managed to reach the source of the urine stream, it would have had to pry open the urethral orifice. But with no appendages to assist in this move, the fish would have been unable to generate sufficient leverage to permit its nearly half-inch-wide head to gain entry.

Spotte also poo-pooed FBC's revised pee-pee claim for much the same reason. "I just don't see how it's possible," he told me. He did believe, however, that it *was* possible that Samad's tale of the strange surgery was true. Primarily, that's because back in 1999, the urologist had shown him a black-and-white video of the candiru-removal procedure. Spotte described what he saw: "The trip to the urethra was like being in the engineer's seat on a subway train, nothing except gray walls all around and ahead, then suddenly the candiru. The alligator clips clasped the tail and started to move in reverse."

Spotte also saw the specimen after its removal, which Samad had preserved in formaldehyde. It measured 5.26 inches in length and had a maximum head width of 0.45 inches.

Beyond questions about FBC's story of the aquatic-turned-aerial assault on his member, there have also been lingering questions about the validity of Samad's video, as well as some of his observations. Of the latter was the physician's claim that the candiru had bitten through the walls of the victim's urethra and into the scrotum. Spotte did find this problematic, given his belief that the candiru's jaws and teeth were specialized for making single bites through the type of feathery tissue associated with fish gills. They would be ill-equipped, therefore, for gnawing through dense tissue like muscle.

Ultimately, though, Spotte believed Samad had removed a candiru from FBC's urethra, but he did not believe the victim's claims about the salmon-mimicking entry of the fish into the site of its last meal. The candiru expert's take was that the trip had been an incredibly rare occurrence, and one in which the underwater entry of the candiru had been completely accidental in nature.

I asked Nathan Lujan to give me *his* take on the FBC story and other claims of rude behavior by candirus. He said that he tended to doubt the account for the same reasons that had been raised by Stephen Spotte. Additionally, there were his personal experiences working with Indigenous peoples throughout Amazonia. According to Lujan, he had "fished in areas where *Vandellia*, the putative culprit in these stories, is relatively abundant, perhaps one per square meter." These were beach habitats that during the day were frequented by bathing children and adults. "In these areas," he told me, "I often ask local residents if they have heard the legend of the candiru, and not once have they independently confirmed any aspect of it."

Lujan added that during a recent trip to Ecuador, he interviewed a Cofán hunter and fisherman who had grown up alongside the river.* "I showed him a picture of the candiru, which he recognized," the ichthyologist said. "I then asked him if he knew any stories about this fish, which drew a blank expression. I explained to him the legend that was common among outsiders. He scratched his chin and said that, no, he had never heard of such a frightening encounter."

The Cofán hunter did tell Lujan that there was a cultural prohibition on urinating in the river. The man didn't know why this prohibition existed but speculated that it might be to avoid encountering candirus.

I asked Lujan what hypothesis he supported regarding instances where candirus *might* have attacked humans? He reminded me that these species had a remarkable ability to nose their way into crevices, and that this ingrained behavioral tendency might explain at least some of the accounts of a candiru mistakenly invading the vagina of a woman sitting, for example, in shallow water.

I decided to conclude our interview by asking the following question: "If I were partially submerged in a section of river known to be a candiru habitat, what do you think the chances would be that one of them would swim up my urethra if I decided to urinate?"

"Zero," he relied.

---

* The Cofán are an Indigenous people who live in northeast Ecuador and southern Colombia, between the Guamués and Aguarico Rivers, both of which contain candirus.

Years before, I'd posed a similar question to Stephen Spotte, this one about the odds of being attacked while partially submerged and peeing in a stream inhabited by candirus.

His answer came quickly. "About the same as being struck by lightning while simultaneously being eaten by a shark."

[3]

Horses: Long in the Tooth

Rather than simplify, researchers, textbook writers, and designers of museum exhibits, to name a few, need to present up-to-date science in an intelligible form, but they should not sacrifice the most modern concepts for the sake of simplicity.

—Bruce J. McFadden

A horse is a horse, of course, of course.

—*Mr. Ed* theme song, by Jay Livingston and Ray Evans

Though its diet is certainly less gruesome than that of the vampire bat or the candiru, the modern horse has also evolved specialized teeth and a suite of related adaptations enabling it to exploit a food source that few other creatures are able to utilize. Here, instead of blood or a chunk of feathery gill filament, the preferred food items are far easier to obtain but far more difficult to digest. The story of the evolution of horses, from small multitoed mammals into hoofed giants, some tipping the scales at over a ton, is another clear example of how shifting environmental conditions can provide the fuel for the engine of evolutionary change—otherwise known as natural selection.

But unlike the hypothesized evolution of vampire bats, this story comes with a rare, well-preserved sequence of transitional fossils from different geological periods, showing anatomical features of both ancestral and descendant groups. Together, this evidence not only provides a fairly complete picture of horse origins but also the successes and failures that took place over the fifty-million-plus-year journey of this surprisingly diverse group. The complexity of this tale also serves as something of a warning about how archaic and oversimplified explanations can lead to a distorted perception of how evolution works.

Horses and their relatives belong to the order Perissodactyla, which contains three extant families: Equidae (horses, asses, and zebras), Rhinocerotidae (rhinos) and Tapiridae (tapirs), plus nearly twenty extinct families. Perissodactyls were also known as odd-toed ungulates, because most of them have feet with either one or three weight-bearing toes (with the remaining digits either absent, vestigial, or pointing backward).* The word "ungulate" is actually an outdated and inconsistently applied term for mammals with hooves, which are, in fact, modified toenails. As with other antiquated forms of classification, researchers no longer use "ungulate" because it has no taxonomic significance—that is, it does not accurately reflect evolutionary relationships. Additional inconsistencies stem from the fact that some ungulates don't even have hooves—like whales and dolphins.

Through the combined results of classical studies in the late nineteenth century and early twentieth century and the work of modern researchers, scientists have been able to uncover what is now known to be a diverse and widespread assemblage of ancient perissodactyls. Like bats and rodents, they were one of many mammal groups that became established by the early part of the Eocene, around fifty-five million years ago. This was roughly ten million years after the extinction of the nonavian dinosaurs and their Mesozoic era pals—many of whom bit the

* Artiodactyls, including pigs, giraffes, hippos, camels, sheep, goats, and cattle, used to be known as even-toed ungulates. They were said to bear their body weight on two toes, with the others either absent, vestigial, or pointing backward.

evolutionary big one some sixty-five million years ago. Their demise not only decreased competition but also left plenty of vacant niches (nature's equivalent of live-in jobs) to fill. For the perissodactyls, their niches centered around consuming plants. American Museum of Natural History curator emeritus Ross MacPhee told me that the first recognizable perissodactyls "were very small, they were forest dwellers, and their teeth were very unlike those of modern horses."

A time traveler to the nineteenth-century would have found a very different form of competition, as paleontologists Othniel Charles Marsh (1831–1899), with Yale University's Peabody Museum, and Edward Drinker Cope (1840–1897), with the Academy of Natural Sciences of Philadelphia, sparred, sabotaged, and discredited each other's efforts at fossil recovery in the American West—a conflict that became known as the Bone Wars.

Looking past their heated rivalry, which lasted from 1872 to 1892,* the pair was integral in bringing about what became a revolution in American paleontology, the study of ancient life. But although their discoveries related to dinosaur fossils garnered more attention from the public, Marsh and Cope were equally important in furthering our understanding of the origins and evolution of birds and horses.

By providing fossil evidence for hundreds of species that were clearly no longer in existence, these rivals also indirectly provided significant support for natural selection, Charles Darwin's recently proposed mechanism by which organisms changed from generation to generation.† Darwin, who decades earlier had teased out the specifics of natural

* The Bone Wars ended in 1892 only because the single-minded quest for paleontological supremacy had left both Marsh and Cope in financial and social ruin.

† Cope actually supported a different theory of evolution, based on the discredited work of Jean-Baptiste Lamarck and others.

selection, was by then a sickly shut-in.* Though scanty by today's fossil record, evidence uncovered by early paleontologists was wielded like a bludgeon by some of Darwin's supporters, notably the English biologist Thomas H. Huxley (1825–1895). Known as "Darwin's bulldog," he used fossils (including ancient human remains) to leave a trail of creationist carnage. For some, the tale of fossil horses would prove to be just as interesting, especially once scientists began to figure out that things weren't quite as simple as they had appeared initially.

Our understanding of horse evolution began with Marsh's 1876 description of the ancient terrier-sized mammal *Eohippus* (Greek for "dawn horse"). Unearthed in Wyoming, the nearly complete skeleton had four toes on each of the front feet and three on the hind feet. Unbeknownst to Marsh, another paleontologist, Richard Owen (1804–1892), had named the same creature *Hyracotherium* nearly four decades earlier. Owen's specimen had been dug out of a seaside cliff near Kent in southeast England. As these things sometimes go, half a century after Marsh named *Eohippus*, a *third* scientist pointed out the double-naming discrepancy. Since Owen had named his fossil first, it had priority, and so *Eohippus* reverted to *Hyracotherium*.† On the bright side, both Marsh and Owen had been dead for over three decades, so nobody's feelings were hurt.

With its short, slender legs, *Hyracotherium* was well adapted for a life hidden on the forest floors of what are now western North America and Europe. Its skull had a short snout, tiny incisors, and low-crowned

* Charles Darwin (1809–1882), who was born on the very same day as Abraham Lincoln, experienced chronic ill health throughout his entire adult life. The identity of his malady has long fascinated and puzzled historians and the scientific community. The list of suspects includes chronic fatigue syndrome, Crohn's disease, cyclic vomiting syndrome, and Chagas disease—and those are just the *C*'s!

† Poor eohippus also lost its italics and capital *E*, since it was no longer a valid scientific name.

(i.e., flattened) cheek teeth (premolars and molars). Taken together, these features indicated that the little creature was a swift runner that fed by browsing on soft plant matter like leaves and fruit. In what stands as the definitive book on the subject, *Fossil Horses: Systematics, Paleobiology, and Evolution of the Family Equidae*, Florida Museum of Natural History paleontologist and leading authority on ancient horse evolution Bruce J. MacFadden suggests that early horses like *Hyracotherium* and its more recent relative *Mesohippus* likely fed on a variety of vegetation, with the up-and-down motion of the jaws and cheek teeth involved in crushing, rather than side-to-side grinding. The latter jaw movement can be seen in modern horses and cows.

As more ancient horse fossils were collected, examined, and named (*Orohippus*, *Parahippus*, *Miohippus*, etc.),* researchers identified a set of adaptations that appeared to transition rather smoothly over geological time, from the Eocene (56–33.9 mya)† and Oligocene (33.9–23 mya) epochs, into the Miocene (23–5.3 mya) and Pliocene (5.3–2.6 mya). The trends that could be seen in the more recent genera like *Merychippus* (in the Middle to Late Miocene) included longer legs and a longer skull, with the eyes situated farther away from the mouth, and, perhaps most significantly, progressively higher-crowned (i.e., longer) teeth (see figure 3).

In the late nineteenth century, Edward Cope noted what he believed to be a linear progression toward increased body size over evolutionary time, a generalization that would become known as Cope's rule. Bruce MacFadden, however, believes this to be an incorrect assessment. He wrote that earlier perissodactyls ranged between 22 and 110 pounds, while more recent species, those living from 20 mya to the present, "were

* Many of these ancient horses also underwent the type of priority-related name change made famous by *Eohippus/Hyracotherium*.

† The abbreviation "mya" = "million years ago."

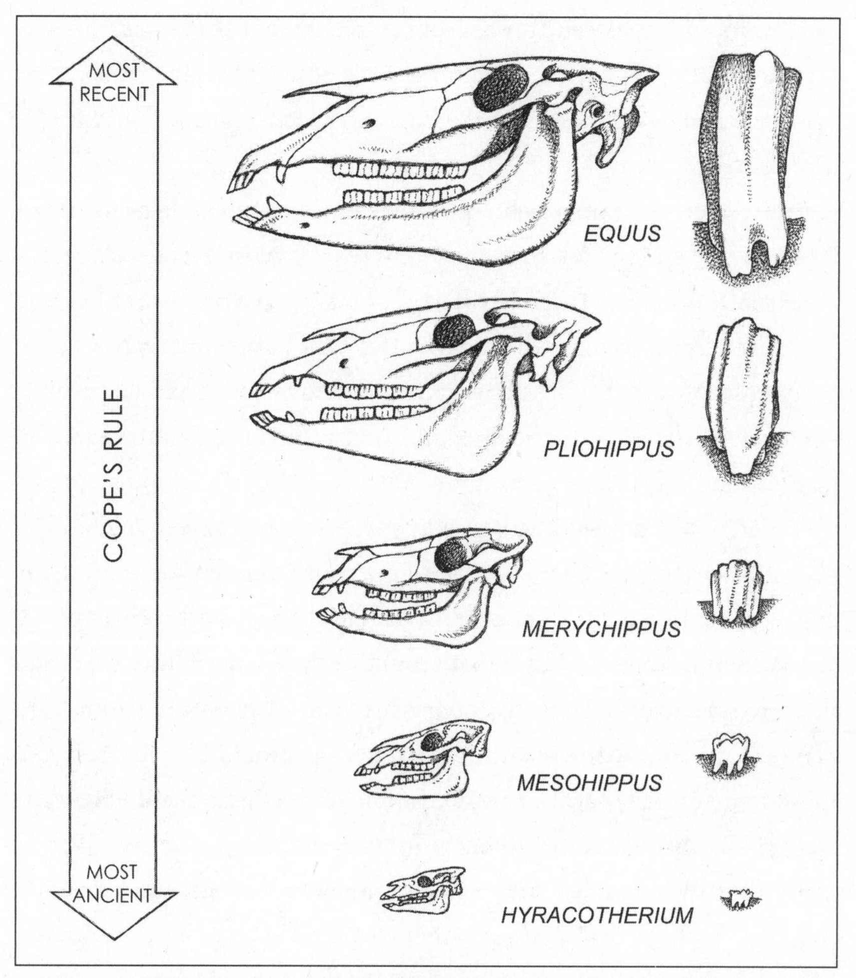

FIGURE 3

more diverse in their body sizes." According to MacFadden, the line that led the modern horses of the genus *Equus* became larger while others stayed the same or even became smaller. Basically, Cope's rule told an evolutionary story that had been oversimplified to the point of inaccuracy.

As twentieth-century fossil collectors examined strata (layers) from the geological periods in question, the reasons for observed anatomical modifications (like a trend toward longer teeth) began to surface. Notably, they saw that changes in the anatomy of the horse fossils they unearthed were matched by differences in the types of ancient plant and seed fossils they also recovered.

More recently came techniques that enabled researchers to detect which of two specific forms of carbon were present in the layers of earth that held the fossils. These isotopes, C_3 and C_4, are important because they can tell researchers which of the two types of photosynthesis were taking place when the fossils were initially formed. C_3 photosynthesis is characteristic of trees, bushes, and herbaceous plants—the types of plants found in forests with an abundant water supply. C_4 photosynthesis is carried out by grasses and grasslike sedges—plants able to tolerate the cooler, drier climates characteristic of open environments like savannas and steppes. In North America, measurable changes in the ratios of the two carbon isotopes indicate that around twenty-five million years ago there was a major shift in the dominant form of plant life (from leafy plants to grasses) with concurrent changes in climate. The key result of these changes was that the massive forests of the Eocene and Oligocene were gradually replaced by Miocene grasslands.

Within this complex story of environmental change, the transition from low- to high-crowned teeth in ancient horses is as enlightening as it is interesting. At its core, the shift in cheek-tooth length is thought to have occurred because of silica (i.e., silicon dioxide), an abrasive compound composed of oxygen and silicon, the two most abundant elements in Earth's crust. Not only is silica an ingredient in glass (which is roughly 70 percent silica, depending on the type of glass), but it is a prime structural component of plant cell walls—particularly grasses, where it can make up 10 percent of dry weight. Another characteristic of grasses is that when compared to leafy greens (the kind that might be eaten by a forest-dwelling browser like *Hyracotherium*), grass contains very little nutritional value. To make a living feeding on it, large amounts of grass must be consumed on a nearly constant basis.

Facing a diet composed primarily of grasses, herbivores with low-crowned teeth (also called brachydont dentition) faced a serious problem: Their teeth would be quickly worn to stumps by the silica-rich grass. And with no chance of growing a new set, they could starve to death and eventually go extinct. This issue is confirmed by the fossil record, where there's ample evidence showing that as environments gradually transitioned from forests to grasslands, species with low-crowned dentition experienced an overall decrease in diversity.

Concurrently, the evolution of high-crowned teeth (also known as hypsodont dentition) in other species contributed to their survival during the forests-to-grasslands changeover. One key reason was that although these teeth were under constant wear and tear, from the abrasive silica-containing grasses being consumed, they continued to grow over the life span of the animal. In fact, unlike the low-crowned teeth found in humans, most of the enamel-covered crowns in horses lie tucked away within the alveoli (tooth sockets), emerging slowly as the exposed crowns are gradually worn down.

Eventually, some researchers felt that attributing the evolution of hypsodont dentition solely to a diet of silica-rich grass was far too simple an explanation. They hypothesized that the near-constant grazing required to survive on a diet of nutrient-poor grass also exposed the grazers to large amounts of sand and soil—or what has been termed "exogenous grit." They proposed that this stony material, adhering to the grass as it was chewed, should not be discounted as a possible driving force in the evolution of high-crowned teeth in the equids (the family containing modern horses). Presumably, the presence of exogenous grit was also a key factor in the independent evolution of high-crowned cheek teeth in other herbivores like cows and deer, as well as in many extinct groups.

The latter included glyptodonts, automobile-sized mammals famous for a dome-shaped carapace reminiscent of a turtle shell.

Weighing in at upward of half a ton, the glyptodont's body armor was composed of hundreds of hexagonal plates of bone. Glyptodonts crossed north into North America around three million years ago, when a land bridge formed in what is now Panama. Like horses and other grass-munching mammals, they had appropriately long, grass-grazing cheek teeth and their skulls had plenty of flat surface area for the attachment of huge jaw muscles. Though glyptodonts had no incisors or canine teeth, some species dissuaded predators with a tail whose tip was equipped with a set of menacing mace-like spikes.

There were further changes to the dentition of ancient horses that were also related to changes in diet. Modern horses and their more recent ancestors have cheek teeth with complex, crescent-moon-shaped cusps. These are the bumps, points, and ridges found on the surfaces of teeth that come into contact with those of the opposing jaw when the jaws occlude (close), the so-called occlusal surfaces. Here, the cusps are indicative of tools used to slice up mouthfuls of tough plant matter. As such, cusp shape can provide information about the diet of its owner, whether that creature is extinct or extant.

On the opposite side of the dietary scale, in carnivores like dogs and cats, some cusps take on an additional function by shearing past each other as the jaws close, like the blades of scissors, slicing off sections of flesh as they do. The teeth involved are the upper fourth premolar (PM4) and lower first molar (M1), here known as carnassial teeth (see figure 4).*

For horses and their more recent ancestors, another adaptation related to a grass-grazing diet was the ability of the mandible (lower jaw)

---

* Teeth are numbered in ascending order, from the front to the back of the mouth. (See figure 4.)

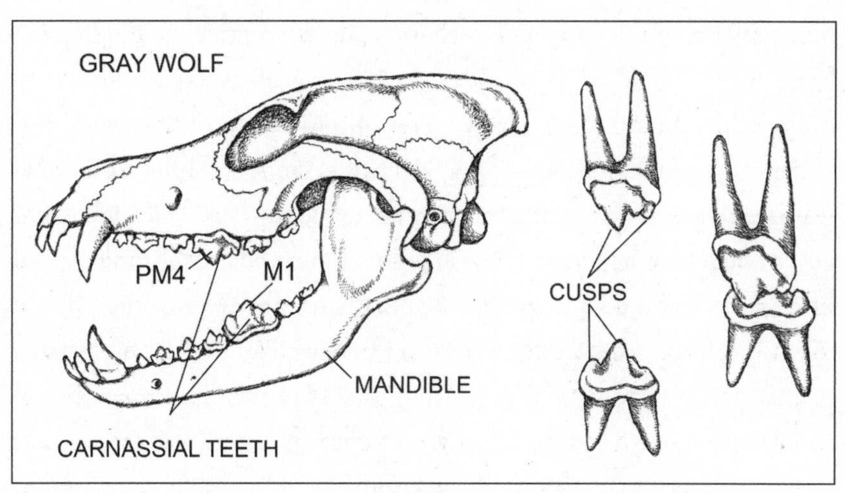

FIGURE 4

to produce the characteristic lateral (side-to-side) movement necessary to efficiently process tough plant material. The mandible also experienced a serious expansion in size, with large, flat surfaces for the attachment of powerful muscles primarily involved in closing the jaw. These features can be easily seen in modern horses and contribute significantly to the muscle-bound look that characterizes the "cheek" region of a horse's head (see figure 3).

In most mammals, adult teeth emerge pretty much full-sized—and except for wear and tear, they generally stay that way. This runs contrary to the adage that as humans get older, they become "long in the tooth." But our teeth aren't elongating with age. Instead, the gums surrounding them are receding, making the teeth appear longer. Causes of gum recession include periodontal disease, aggressive brushing, changes in hormone levels, genetics, and smoking. It is quite likely that the term "long in the tooth" originated with horse owners. That's because, as mentioned earlier, horse incisors *do* grow during adulthood, and so tooth length can serve as an indicator of equine age. The expression "Don't look a gift horse in the mouth" appears to share a similar origin, the implication being that it would be an insult to the gift giver if the recipient were to examine the mouth of a horse given as a gift to determine its age.*

Beyond adaptations related to dentition, horses exhibit additional features related to changing environmental conditions. With less plant cover under which to hide, being able to escape predators through speed likely replaced the adaptations for stealth often seen in low-to-the-ground forest dwellers. From a biomechanical perspective (that is, by looking at anatomical structures as machines), there are two ways to get faster: (1) increase stride frequency (i.e., move your legs faster), and (2) increase stride length, so that each time your limbs move forward, more ground is covered.

For horses, the loss of digits left them walking/running on the tips of their middle toes, an evolutionary trend that lengthened their legs *and* their stride. In one of many examples of convergent evolution, this adaptation was integral in generating increased speed in many unrelated species (like antelopes and even ostriches) that also live on open, grass-covered terrain.†

One more adaptation to a grassland environment for horses was the repositioning of the eyes farther away from the mouth—which was itself located at the end of a snout that was becoming progressively longer. (See figure 3.) These features allowed the grazers to keep a sharp lookout for predators while feeding, which had become something of a full-time job.

---

* While some other tooth-related idioms make sense (e.g., "fighting tooth and nail" and "red in tooth and claw"), others are less clear. "Clean as a hound's tooth" remains a puzzle. Why would hound's teeth be any cleaner than those of a terrier or, for that matter, any other canine?

† Convergent evolution (also known as evolutionary convergence) occurs when a particular feature (like wings) is shared by organisms that aren't closely related (like birds and flying insects) and was *not* passed down to them from a common ancestor. Instead, that feature evolved separately in each group as the organisms adapted to similar environments or niches. The fusiform bodies (i.e., tapered at both ends) of sharks (fish) and dolphins (mammals) are another example, as are the long legs seen in runners, like horses (mammals) and ostriches (birds).

There were also important changes to the anatomy of horse ancestors that were not preserved in the fossil record. This is due to the fact that they occurred in soft tissue structures like organs.

The bottom line is that leaves and tender plant bits are relatively easier to digest than grass. They also contain more nutrients than the same volume of grass. As a result, grazers are characterized by having digestive tracts (stomachs and intestines) that are larger and longer than those of forest browsers. The idea here is: the more time that the food remains in the digestive tract, the more nutrition that can be derived from it. This is a hallmark of animals in which sturdy teeth and digestive enzymes are not sufficient to free up the sparse nutrients locked in material like grass.

In horses and their relatives, this meant the evolution of longer intestines, equipped with an enormous pouch-like extension called a cecum.* Other grazers, like cows and their kin, have enlarged stomachs divided into separate compartments (like the rumen), through which the food moves sequentially.

But it's not just intestine length or stomach volume that increases the efficiency of digestion in grazers. Both of these organs are full of helpful gut bacteria. These so-called endosymbionts get a warm place to live, while providing their cosymbionts (the horse or cow) with an ability no multicellular organism, whether vertebrate or invertebrate, can handle by itself—namely, breaking down cellulose into nutrients that their bodies can use. Cellulose is a polysaccharide, a very long chain of sugar molecules (in this case glucose) that is the sturdy, primary component of plant cell walls. Since cell membranes (the animal equivalent of plant cell walls) are not composed of cellulose, this explains why meat is easier to digest than plants (and why carnivore digestive tracts are relatively

* The now-functionless remnant of the cecum in humans is the vermiform appendix, which points to a more plant-based diet in our own ancestors.

shorter). It also explains why herbivores chew their food longer, crushing the plant cell walls to release the nutrients found within their cells. Once the well-chewed food moves into expanded regions of the gut, the microbes involved in cellulose digestion use the chemical process of fermentation to convert cellulose into glucose, which *can* be further digested by their hosts—a list that even includes wood-chomping termites.*

As envisioned by late-nineteenth-century paleontologists, the progression of fossil perissodactyls was depicted in a straight line based on some combination of increasing body size, tooth length, and skull length, and a decreasing number of toes. This reflected a hypothesis that became known as orthogenesis, basically the belief that evolution worked in a stepwise, linear fashion, with each ancient species on a march toward progress and some evolutionary destiny.

Though Edward Cope eventually came to believe in evolution, like others of his time he did not believe that it was a random process. These folks rejected Darwin's concept of natural selection, contending instead that evolution was goal-oriented. As proposed by Darwin, the mutations that provided variation in a population were chance occurrences—not directed toward some ultimate goal. A few of these mutations, though, like those that produced longer teeth in horses, enabled some ancient species to survive significant environmental changes (like the transition from forests to grit-filled grasslands) while those with shorter teeth died out. Passed on to future generations, longer teeth would then be considered adaptations.

Because nobody (including Darwin) quite knew how these variations showed up or, especially, how they were passed on to the next generation,

---

* In addition to perissodactyls, other so-called hindgut fermenters (those with a cecum) include rabbits, elephants, and many rodents. Foregut fermenters (those with modifications to the stomach and/or esophagus) include cattle, goats, sheep, antelopes, deer, and giraffes.

orthogenetic phylogenies became an extremely popular way to illustrate human evolution as a sort of march toward perfection.[*] But whether the parade consisted of horses or humans, the march ended with the modern form standing at the front of a line or atop a column of imperfect losers, each destined for replacement and extinction. Various explanations for how this occurred (orthogenesis, theistic evolution, and Lamarckism)[†] were bandied about, tweaked, and eventually abandoned during the latter part of the early twentieth century.

By the 1930s, the work of Charles Darwin, who had been dead for half a century, was being reexamined in the light of Mendelian genetics. This was the so-called modern synthesis. Gregor Mendel's groundbreaking work, published in 1866, which explained how variation in pea plants (*Pisum sativum*) could be passed on to future generations, was unfortunately not read by Darwin[‡] (or many others, for that matter). But within a decade of the paper's twentieth-century rediscovery, a series of influential publications from a cadre of high-powered science types strongly supported natural selection as the mechanism that drove evolution.

Paleontologists were also collecting, naming, and classifying more and more fossil species, and eventually straight-line phylogenies were replaced by something that more resembled a bush or a tree. (See figure 5.)[§] In these representations, some side branches would come to an end,

---

* Phylogenies, in this sense, are graphical representations of the evolutionary history of a group of organisms.

† Jean-Baptiste Lamarck (1744–1829) was the first naturalist to propose a mechanism for evolutionary change. He correctly hypothesized that environmental change was the stimulus for evolution. Unfortunately, he is remembered best for his belief that the needs and wants of the individual led to the development of adaptations (like longer necks in giraffes). This concept, known as "the inheritance of acquired characteristics" wasn't even his, but it turned Lamarck into a convenient laughingstock in numerous biology textbooks and in popular science literature.

‡ Although a sealed copy of Mendel's paper was found among Darwin's belongings after his death in 1882.

§ Adapted from MacFadden, "Fossil Horses."

indicating lineages that had gone extinct. Other branches might continue unbroken into current times (the region of the tree farthest from the base), representing extant forms. By simultaneously incorporating what was being learned about major environmental changes during Earth's history, these branches could all be explained by the process of natural selection—now strongly supported by genetics and other scientific fields like geology.

With little competition among the ancient horses for the tough, nutritionally deficient grasses (and the sandy particles that adhered to them), several lines of ancestral perissodactyls thrived and diversified, some moving into the Old World but most inhabiting North America. There, they were wildly successful, with a dozen or more species living concurrently around ten million years ago. Contrary to Cope's "bigger is better" mantra, some of these new- and old-world species were smaller-bodied or even dwarf forms, while others bucked the simplistic concept of orthogenesis by reverting to mixed diets of grasses and leafy plants. Eventually, though, like 99.9 percent of all the species that have ever existed on Earth, these ancient mammals went extinct, most leaving no modern descendants. What did survive into modern times was a single genus, *Equus*, which currently contains around ten species. These include horses, donkeys, asses, and three types of zebras.

As for the fate of the modern horse in North America, Paleo-Americans and climate change ultimately drove many large mammals to extinction around ten thousand years ago. That included the North American *Equus* species, but not before some of them had crossed the Bering Land Bridge, which formerly linked what is now Alaska and Siberia. These new arrivals diversified as well, spreading into Asia, Europe, and Africa, and some of them were eventually domesticated by the humans who lived there. Somewhere around six thousand years ago, members of the lineage that would become modern horses (*Equus caballus*) were domesticated from horses like the wild Przewalski's

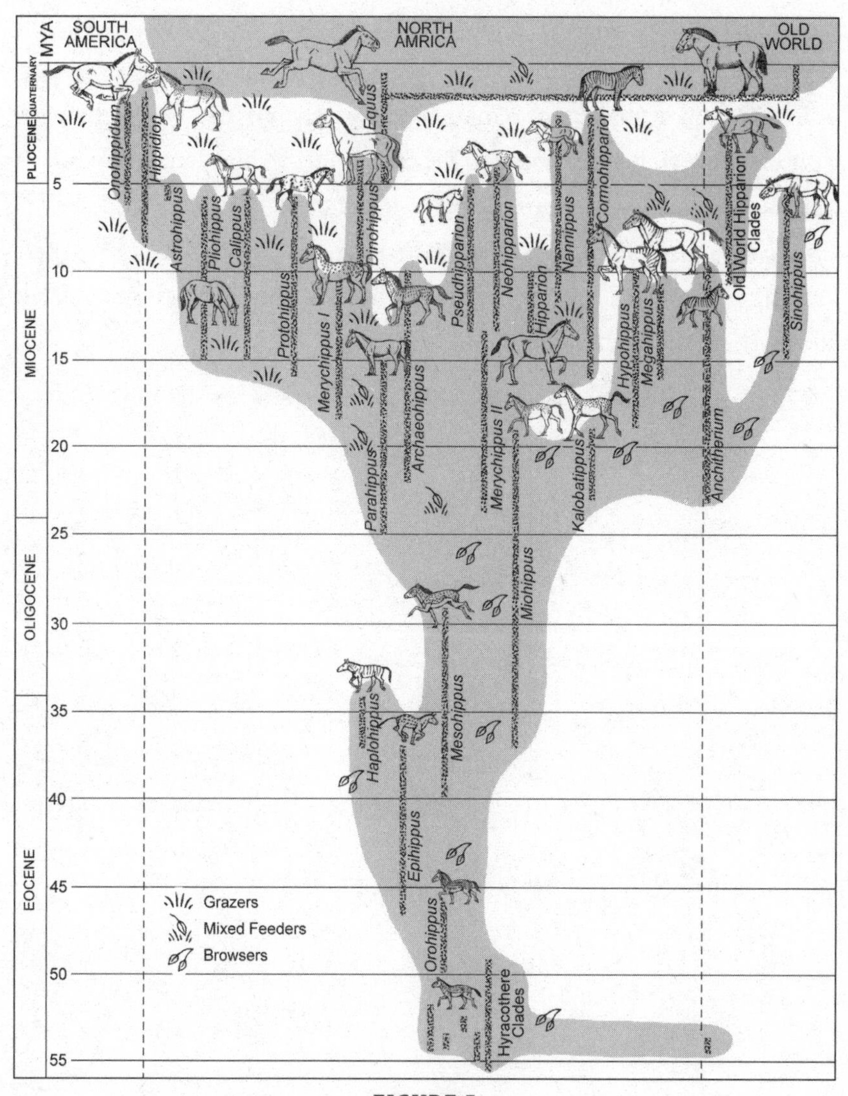

FIGURE 5

horse, native to the steppes of Central Asia. These were introduced (though some might say reintroduced) into the New World by the Spanish during the late fifteenth century and the early sixteenth century, still beautifully adapted to the vast grasslands of North America that their recent ancestors had once roamed.

Unfortunately, since it is admittedly simpler to consider a straight line than it is to ponder the intricacies of a bush, the stepwise evolution of horses (and humans too) remained extremely popular, especially with a public that had been exposed to the concept through a long list of books and natural history museum displays, some of which still exist. Even so, the success of modern horses, and the long, strange trip taken by their ancient and modern relatives, is one of the most interesting tooth-related stories in all the animal kingdom.

[4]

The Tusked and the Tuskless

I wish I was a narwhal so that I could just stab idiots with my head.

—Unknown

Roughly 270 million years ago, before the Age of Dinosaurs, and around 60 million years before the first mammals, a diverse assemblage of vertebrates known as synapsids dominated the landscape. They differed from the two other vertebrate groups in the number of fenestrae (or openings) found on each side of their skull. The anapsids (like modern turtles and their ancestors) had none of these fenestrae, while the diapsids (lizards, snakes, crocs, pterosaurs and dinosaurs) had two. Synapsids, which would one day include horses and insurance salesmen among their ranks, had a single temporal opening on each side. Through it, the fan-shaped temporalis muscle passes from the side of the skull, across the jaw joint, and onto the mandible. Thus situated, the temporalis and the similarly jaw-joint-spanning masseter muscle provide the energy for a bite, as they contract and draw the jaws together (see figure 6).

This example clearly provides a response to those who might question how a paleontologist can tell a great deal about an animal by looking at a single bone. If the bone happens to be a mandible with a broad surface area, the researchers knows that the animal had large muscles attached to that surface and thus a powerful bite. A delicately

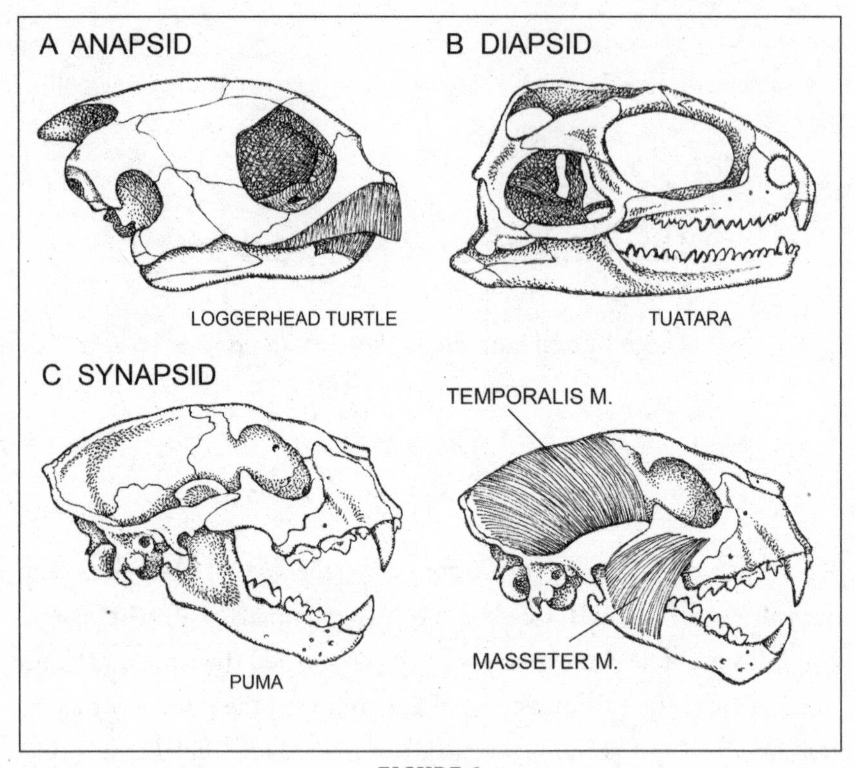

FIGURE 6

built mandible, such as that found in a hummingbird, would indicate a weak bite.

But no matter the number of temporal fenestrae or how powerful a bite these critters may have exhibited, most of the ancient synapsids met their demise during the greatest extinction event the planet has ever experienced. This was not however, the end-Cretaceous extinction, the one that killed off everyone's favorite kiddie toys roughly 65 million years ago. It was the Permian-Triassic extinction, which took place around 187 million years earlier. Also known as the Great Dying, it may have been the closest that life on this planet has come to disappearing completely, wiping out over 95 percent of marine species and 70 percent of land dwellers. Unlike the asteroid-fueled end-Cretaceous extinction, the causes of the Permian-Triassic event are still being debated, though they likely involved massive volcanic activity and a warming climate. Luckily for us, one group of synapsids that survived the catastrophic die-off was the mammal-like cynodonts, and some of them eventually evolved enough mammalian characteristics that two hundred million years later their bipedal descendants (i.e., humans) stopped calling them mammal-like reptiles (or reptile-like mammals) and started referring to them as mammals of the card-carrying variety.

Another group of synapsids that survived the Permian-Triassic extinction were the dicynodonts ("two" + "canine" + "teeth"). These herbivores were themselves descended from the group that gave rise to the sail-backed reptiles of the Early Permian period (295–272 mya). These were popularized by the mistaken inclusion of the sail-backed carnivore *Dimetrodon* in dinosaur toy collections. But while the dicynodonts also included some admittedly weird members (the mole-like burrowers *Cistecephalus* and *Diictodon* come to mind), the reason they're mentioned here is that they were the first known vertebrates to evolve tusks—structures otherwise confined to mammals.

Tusks are, in fact, teeth, but they're not used for mastication (i.e., chewing). Instead, they're employed as digging or scraping tools, for visual display, and, in some species, for aggressive interactions. Walruses also use their tusks to haul their massive bodies out of the water and onto the sea ice ("tooth walking") and to maintain the breathing holes they've created by poking through the ice from below.

Evidence indicates that, as they would in the mammals, tusks evolved on multiple occasions in the dicynodonts, in this case with a gradual transition from oversized canines solidly attached to the jaw to true tusks with a more fibrous attachment to the jaw by way of the same type of periodontal ligaments that currently hold our teeth in place. In the case of tusks, though, the arrangement allowed for continuous growth.

Combined with their ability to powerfully process plant material, these enlarged, deep-rooted upper canines were key innovations that likely contributed to the success and survival of the dicynodonts (through the end-Permian extinction) for some sixty-four million years—until they were finally driven to extinction in the Late Triassic period.

Later dicynodonts (like *Dolichuranus*, which should translate to "Why I'm Never Mentioned in Children's Books") kept their tusks but lost the remainder of their teeth, a loss compensated for by the evolution of turtle-like beaks. With strong limbs, useful for digging but certainly not running, some of these lumbering creatures dealt with predators (for example, the wolflike *Cynognathus*) by attaining elephantine size (see figure 7).

FIGURE 7

Since the demise of the dicynodonts, only mammals possess tusks, which are defined as ever-growing incisors or canines. Their most obvious characteristic is that a portion of the tusk protrudes from the oral cavity. The indeterminant (unlimited) growth of these structures differentiates them from other elongated mammalian dental structures that were of determinant growth (i.e., ceasing during adulthood), like the teeth of *Smilodon* and other now-extinct saber-toothed cats.

Tusks, like all teeth, are primarily composed of dentin. One difference is that enamel is not usually present on the sections of the tusk that emerge during adulthood, often leaving a cap-like section of enamel on the tusk tip that eventually wears away. This distinguishes tusks from other ever-growing teeth, like the incisors of rodents, where dentin *and* enamel are constantly being laid down as more and more of the crown erupts, becoming visible through the gums. Note that the chisel-like nature of rodent incisors, perfect for cutting through tough, fibrous plants, is made possible by the fact that enamel is deposited only on the labial (lip) side of these teeth and not the lingual (tongue) side. Consequently, the enamel-free surface of the incisor wears down quicker, with the more durable enamel-coated surface forming a longer, sharpened edge (see figure 8).

As they did in some nonmammalian synapsids, tusks evolved multiple times in the mammals, with the list of living species composed of elephants (the largest living tusked mammals), hippopotamuses, walruses, manatee-like dugongs, muntjacs (picture a 2.5-foot-tall deer impersonating a Hollywood-style vampire), wild boars, warthogs, narwhals, and hyraxes (the smallest tusked animals). My personal favorite is the babirusa (*Babyrousa* spp.*), one of several endangered pigs found on the Indonesian island of Sulawesi. They're equipped with backward-curving

---

* The abbreviation "spp." is the plural form of "sp."—which stands for "species."

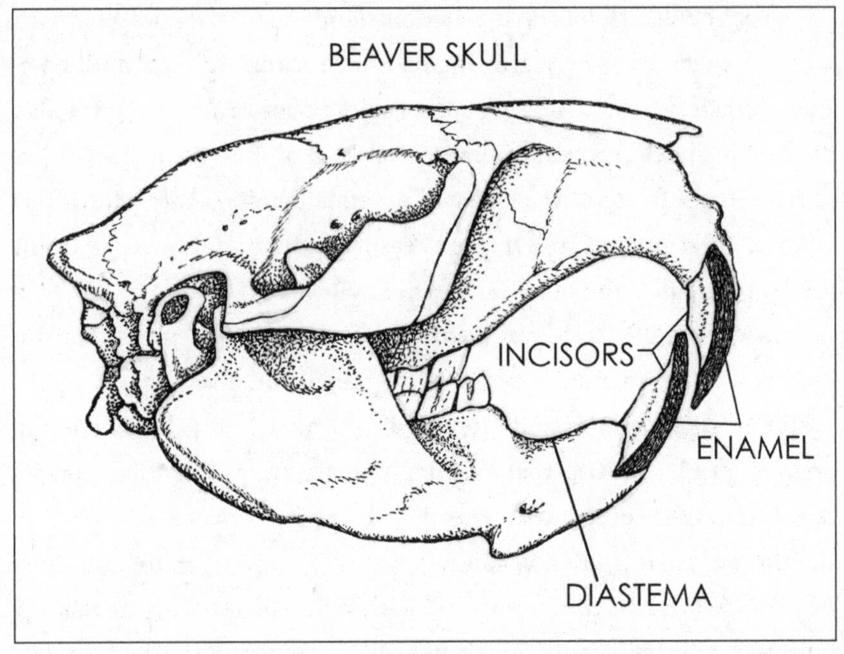

FIGURE 8

tusks, whose function is still under debate; in rare instances, these structures can grow until they pierce the skull of older babirusas, killing them (see figure 9).

In the 1570s, Sir Martin Frobisher, former privateer and newly minted English explorer, led three expeditions to the Canadian Arctic. His goal was to find and chart a northwest passage that would eliminate the considerable hassle of sailing around the tips of South America or Africa to reach the riches of Asia. When his first voyage brought him to within a mere four thousand miles of his goal (at the cost of a ship and the lives of some of his men), Frobisher switched gears. As was becoming customary for European flag planters, he put finding a shorter route to Asia on the back burner and began searching for gold. Closely following the "Gold at Any Cost Handbook," which would have been required reading for explorers from Christopher Columbus onward had it actually existed, Frobisher clashed with the locals—in this case, the Inuit people, who were quickly accused of cannibalism and witchcraft, oft-repeated claims among Europeans looking for the shiny stuff. Frobisher's men reportedly

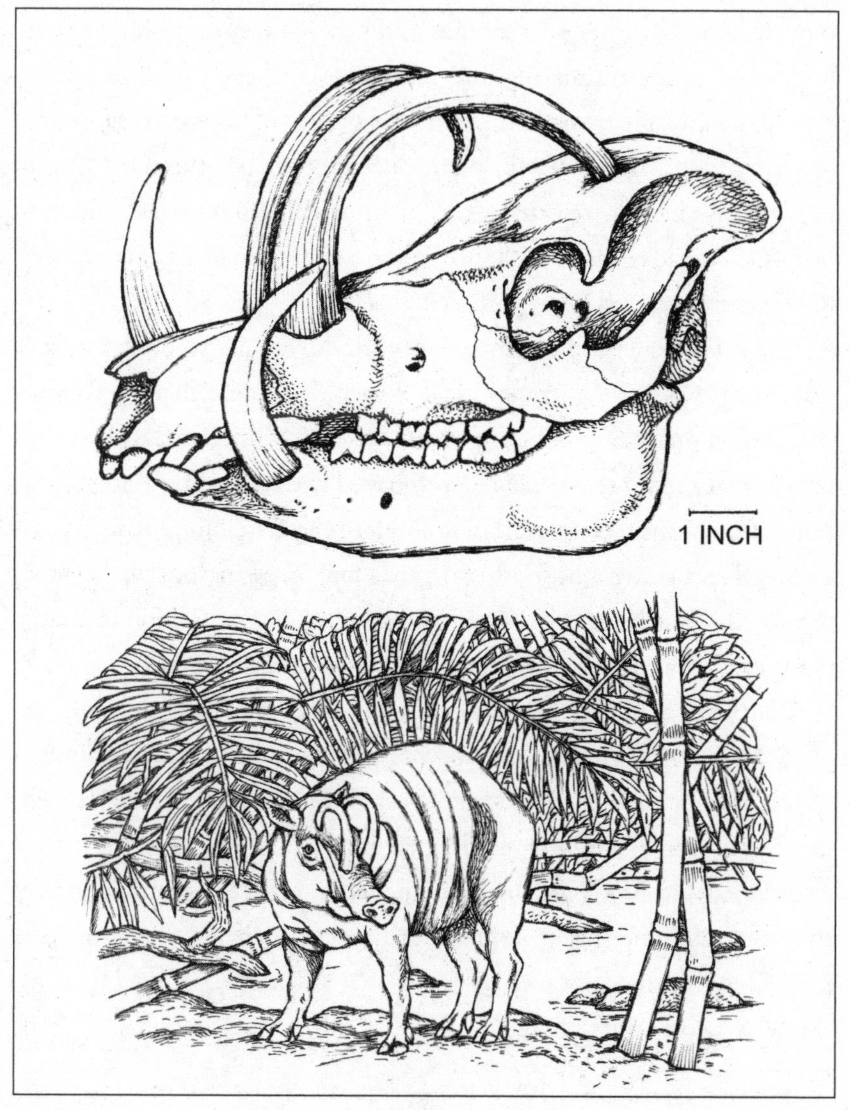

FIGURE 9

went so far as to check the feet of the locals they encountered for cloven hooves.

Sticking to what had worked for other treasure hunters in the Caribbean, Mexico, and South America, once the Inuit had been suitably dehumanized Frobisher and his men treated them as pests. They reportedly opened fire on a mother and her infant (wounding the suckling baby), while murdering and kidnapping others at will. Frobisher eventually took an arrow in the butt, and several of his crew disappeared or were killed. Ultimately, his men collected over one thousand tons of ore, but his smelting operation turned up nothing but useless minerals. Frobisher wrote of his haul, "It proved no better than blacklead, and verified the proverb: All is not gold that glistereth."

What Frobisher did discover, however, during his second voyage in 1577, was the body of a strange creature, which he described in his journal: "Upon another small island here was also found a great dead fish, which, as it should seem, had been embayed with ice, and was in proportion round like to a porpoise, being about twelve foot long, and in bigness answerable, having a horn of two yards long growing out of the snout or nostrils. This horn is wreathed and straight, like in fashion to a taper made of wax, and may truly be thought to be the sea-unicorn."

But this was no fish. It was a mammal, a narwhal (*Monodon monoceros*). Like other whales, the narwhal is believed to share a common terrestrial ancestor with the hippopotamus. Its classification has also undergone some shuffling, as old groupings (like the ungulates) were abandoned for new ones thought to better reflect evolutionary relationships. Numbering around ninety species, whales, dolphins, and porpoises have been placed in their own infraorder,* Cetacea, which contains two major subdivisions: the Mysticeti (baleen whales), to be

* "Infraorder" is a taxonomic rank below "suborder" (which sits below "order") and above "superfamily" (which sits above "family"). And if this sounds confusing—it is.

discussed in chapter 13, and the Odontoceti (literally, "toothed whales"), which includes narwhals and their close relatives the beluga whales (*Delphinapterus leucas*).

Odontocetes, which include dolphins, porpoises, and the fictional sperm whale Moby Dick, typically have a single row of teeth that are roughly the same shape and size, a dental condition known as homodonty (from the ancient Greek for "same teeth"). With the exception of armadillos, this feature sets the toothed whales apart from the rest of the mammals. Most of those (including humans) exhibit heterodont dentition, where a mouth can contain up to four different types of teeth. Across the vertebrates, though, homodont dentition is the norm, occurring in most fish, amphibians, and reptiles. But although narwal dentition has been described as homodont, in reality it is something altogether unique.

Characteristically, but not always, narwhal tusks are single and spirally twisted along their lengths, which can reach up to ten feet. This twist always runs in a counterclockwise direction when viewed from the base to the tip.

Narwhal tusks are modified upper canines that emerge from the maxillary bone (or maxilla), the principal bone of the upper jaw. In addition to giving rise to the upper canines, premolars, and molars, the maxilla forms the hard palate and the lower part of the orbits (which contain and protect the eyes) and makes up much of the face. The premaxillary bone sits directly in front of the maxilla, where it forms the tip of the upper jaw and gives rise to the upper incisors. In elephants, two of these incisors can grow into tusks.

When dental surgeon and marine mammal specialist Martin Nweeia and his colleagues examined 131 narwal skulls, most of them museum specimens, they proved that narwhal tusks are indeed canine teeth. They also examined the anatomy of narwal maxillary bones and the teeth that emerge from them and determined that there are six pairs of fetal tooth buds on the upper jaw, though only two of those pairs develop any further. In males, the front-most maxillary tooth buds develop into tusks but typically only the one on the left side erupts, piercing the upper lip and growing over the entire life of the individual. The second set of tooth buds (located just posterior to the first) forms tiny vestigial teeth, usually less than six millimeters in length by three millimeters in width. In most females, both tusks remain unerupted. In rare cases, both tusks emerge in male narwhals and one tusk erupts in females.

I contacted Royal Ontario Museum mammal specialist Jacqueline Miller to get some additional information on narwhals, especially as related to their tusks. Miller and her ROM colleagues have a serious interest in whale species inhabiting their local waters, and so narwhals, with populations in northern Hudson Bay and Baffin Bay (as well as the Greenland Sea and the Arctic Ocean) fit the bill perfectly.*

"Narwhals are sexually dimorphic," Miller told me. "Not only in the possession of tusks, but in body size as well." Sexual dimorphism is a condition where there are significant and visible differences between males and females of the same species. The differing plumage of peacocks and peahens are a well-known example.

Miller also pointed out that there is a strong correlation between the size of the male and the length of its tusk. "It's because of this body size/tusk length relationship that the evolution of the narwhal tusk has been hypothesized as under strong sexual selection."

I nodded. "So . . . bigger tusk equals larger-sized male . . . equals increased desirability in the eyes of a female."

Think of the showy displays by male peacocks during the mating season. While we might consider the quivering of brilliantly colored six-foot feathers (which emits low-frequency sounds inaudible to human ears) to

---

* The five-year-long effort of Jacqueline Miller, Mark Engstrom, Burton Lim, and their fellow ROM colleagues to preserve a blue whale heart for the first time were documented in my book *Pump: A Natural History of the Heart.*

be overkill, from a peahen's perspective all this theater would translate to something like "If he's healthy enough to expend all that energy on bright feathers and attractive sounds (or a long tusk), then he's healthy enough to father my babies."

"Part of the idea of sexual selection involves the tusk serving as a visual proxy of strength and competitiveness," Miller confirmed.

This body size/tusk length relationship may be one of the factors in narwhal behavior known as tusking, during which two males will touch tusks and jostle each other about. Tusking activity may be a way to measure up opponents during courtship—sort of a means of determining who the "winning" male is, without actual fighting.

According to Miller, though, "There are notable observations of scarring around the head of males, as well as broken or damaged tusks, that suggest aggressive interactions can end up in out and out fighting." It turns out, she said, "the tusk can endure significant blows from the side but appears to poorly resist the stress of head-on blunt impact."

In addition to its use in sexual selection and jousting, I had also heard that narwhal tusks might function as a sensory organ. Miller told me that this hypothesis had come about because of the presence of specialized structures identified during dissections of these colossal canines (see figure 10).

In most mammals, cementum is the fibrous material that binds teeth to the jaw. In narwhals, this collagen-rich substance makes up the outermost layer of the tusk, imparting flexibility to the structure. Researchers found tiny channels in the cementum that not only open to the outside environment (i.e., surrounding water) but also connect to minute tubes that course throughout the dentin that makes up the next-deepest layer of the tusk. Located deeper still, and like other mammal teeth, narwhal tusks also have a central pulp cavity containing blood vessels and sensory nerve endings. The respective function of the arteries and veins is to supply the tooth (or in this case, tusk) with oxygen and nutrients and to carry away deoxygenated and waste-laden blood. And as anyone who's ever had an achy or temperature-sensitive tooth has experienced, the sensory nerves receive information from sensory receptors in the tooth (like pain and touch receptors), warning the owner of potentially dangerous changes to the condition of the tooth.

Experiments by Martin Nweeia and his colleagues on narwhal tusks indicated that besides providing information to the brain about tooth condition, this sensory system evolved to send along additional information about environmental changes in the surrounding water. They determined that sensory receptors in the tusk, activated by changes in the salinity of the water as the narwhal swims through it, would stimulate the sensory nerves associated with the receptors. These nerves would then send signals (via nerve impulses) on to the brain by way of the maxillary branch of the trigeminal nerve.

As for why detecting changing salinity gradients might be important, Nweeia and his coauthors hypothesize that this ability might enable male narwhals traveling under the Arctic ice sheet to detect areas of more dilute (i.e., less salty) water that occur near holes where the whales can surface to breathe.* Narwhals, like all mammals, will drown if they cannot surface for air. For males, known to dive deeper than females, this environmental monitoring system might allow for easier access to breathing holes when returning from deep dives.

According to Arctic researcher Cortney Watt, there is much to learn about the iconic narwhal, including its diving and reproductive habits (observations of narwhal mating behavior have never been reported). Additional information related to the still-evolving centuries-long

---

* The reason for the decreased salinity around breathing holes appears to be that sea ice holds less salt than normal seawater, so when it melts around the perimeter of a breathing hole, the water within the hole is less salty.

relationship between the whales and the Inuit who hunt them will also be vital, as will data on the effect of global climate change on narwhal populations and behavior. Studies have already shown that a decrease in Arctic sea ice has enabled killer whales (*Orcinus orca*), narwhals' primary

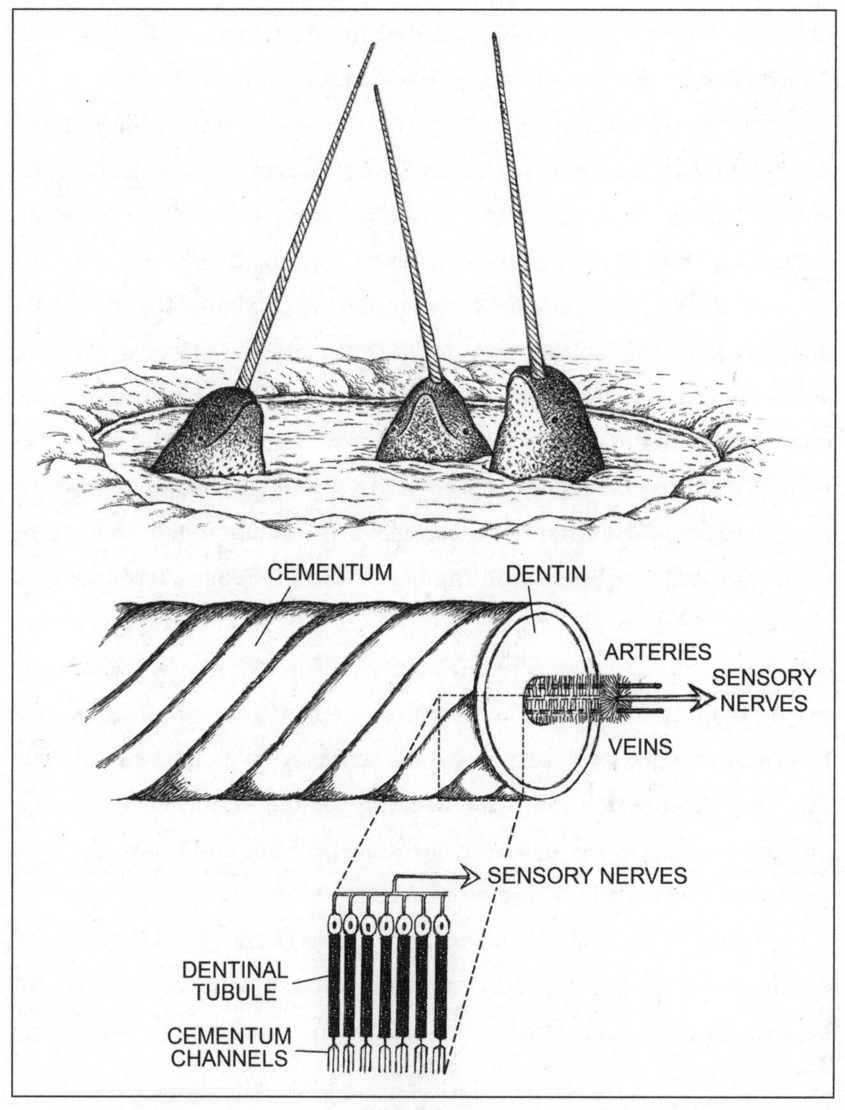

FIGURE 10

nonhuman predator, to expand their range into areas populated by narwhals, a species whose own diets also appear to be changing.

Researchers hypothesize that as narwhals have encountered changing environmental conditions like reduced sea ice, they have adapted their feeding habits to include more open-water fish species like capelin (*Mallotus villosus*). The problem is that these species are likely to have high levels of industrial pollutants like mercury in their tissues.

Scientists remain intrigued by narwhal tusks, especially the possibility that these curious structures may have even more functions than previously known. Evidence from a drone video appears to show narwhals using their tusks to stun Arctic cod before gulping them down.* This has not been definitively confirmed, but researchers do know that narwhals' uniquely elongated canines grow for the entire life of their owner, as new layers of bone are laid down on previous layers. That's because the process leaves growth rings reminiscent of those seen in tree trunks. And as with the dendrochronologists who study tree rings, researchers studying the tusks of dead narwhals can glean information about changing environmental conditions and the health of individual narwhals when each layer was formed. This analysis can also identify the chemicals each narwhal encountered at different points during its life span, which can extend more than fifty years. Alarmingly, recent analyses of tusks have revealed increased levels of mercury—a seriously toxic chemical element. Tied into this observation is the widespread, long-term, and dangerous problem of mercury exposure from industrial pollution, namely coal-fired power plants and incinerators.

Mercury in the water is taken in by bacteria and converted into an organic form known as methylmercury. Microorganisms consume the bacteria, tiny animals (like krill) eat the microorganisms, and larger

* "Arctic cod" is an ambiguous common name of several genera in the cod family Gadidae, related to true cod (*Gadus* spp.). Here, it likely relates to *Arctogadus glacialis*.

animals (like capelin) eat the krill. At each step in what has been termed a toxicity pyramid, the methylmercury becomes more concentrated, in a process known as bioaccumulation. After feeding on the capelin, narwhals are exposed to the highest concentrations of mercury. Thus, they sit (albeit temporarily) at the capstone of this particular pyramid.

According to Aarhus University Arctic researcher Rune Dietz, "Whales lack the physiological properties to eliminate environmental contaminants. They don't get rid of mercury by forming hair and feathers like polar bears, seals, and seabirds, just as their enzyme system is less efficient at breaking down organic pollutants."

Mercury also poses a real danger to the Inuit, who, in addition to their diet of fish, have long-consumed narwhals and beluga whales. Because of this, these Indigenous people now stand above the whales at the pinnacle of the toxicity pyramid. Recent studies have shown that Inuit are exposed to mercury at levels that are among the highest in the world—levels that are causing neurodevelopmental problems (like lower IQs) in children with prenatal exposure to the toxin. In places like northern Canada, dietary counseling programs are attempting to educate Indigenous populations about health concerns, including fetal sensitivity to mercury exposure.

UNDOUBTABLY, THE MOST popular tusks in the animal kingdom belong to the three extant species of elephants, the African bush elephant (*Loxodonta africana*), the forest elephant (*Loxodonta cyclotis*), and the Asian elephant (*Elephas maximus*). Composed primarily of dentin, elephant tusks are modified upper incisors—which differentiates them from the elongate left upper canines of narwhals.*

But it's not just elephant tusks that make elephant dentition so

* Elephants do not have lower incisors.

interesting. Elephant teeth are not replaced vertically, as in most other mammals, including humans. Instead, their molariform teeth (premolars and molars, labeled in figure 11 as M1 through M6) move forward, conveyor-belt-style, with the worn-out anterior-most teeth dropping off to be replaced from behind. This is the case for manatees too, though they possess no front teeth at all—only molars.

Most mammals have but two sets of teeth over the life of the individual, a dental condition known as diphyodonty. In these creatures, there's a set of deciduous teeth (also called primary teeth, baby teeth, or milk teeth) and a set of permanent (or adult) teeth to replace the deciduous teeth when they fall out. Things are quite different for elephants and manatees.

Over their long lives, elephants will go through a *total* of six molariform teeth—three premolars (M1–M3) and three molars (M4–M6) on one side of the upper or lower jaw. An old elephant may have only a single molar (M6) on each side of the upper or lower jaw. It will die of starvation when those teeth wear down.

Lacking incisors, canines, and adult premolars, manatees (*Trichechus* sp.) exhibit a similarly strange dental phenomenon. Rather than deciduous teeth being replaced vertically, with an adult tooth waiting to grow into the spot where the baby tooth fell out, manatee tooth replacement is horizontal, as in elephants. Once these so-called marching molars reach their final position (through a process known as hind molar progression), they're used to grind down the aquatic plants that make up the manatee diet. Eventually, these frontline molars are worn down by sand and tough plant matter, at which point their roots are resorbed and they fall out—to be replaced by the next molars in line. The number of teeth that manatees have over their lifetime varies with how coarse their diet is and the sand content of their feeding areas.

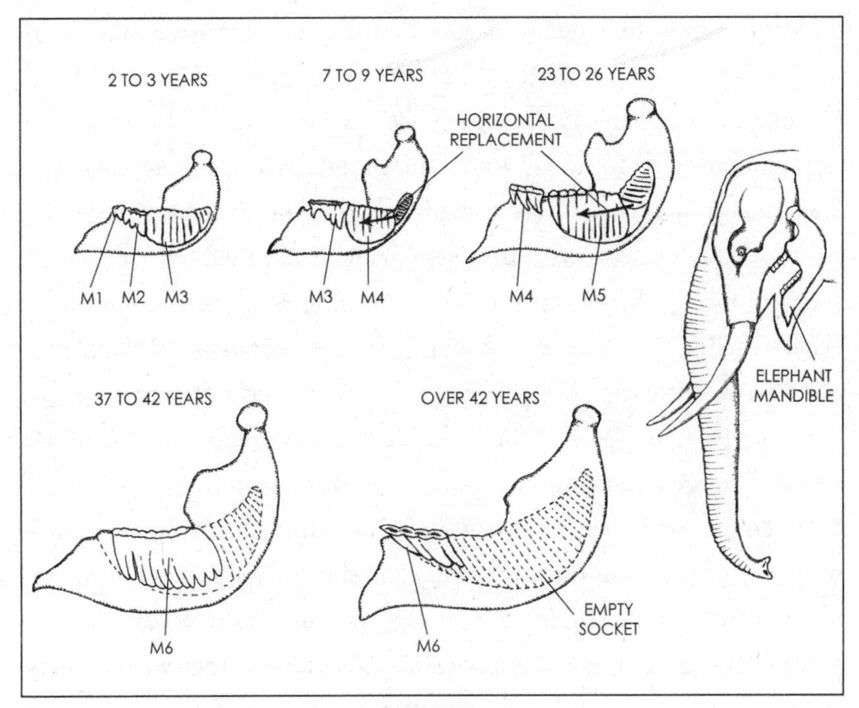

FIGURE 11

As for elephant tusks, approximately two-thirds of these jumbo incisors is visible, while the remainder sits in a socket in the skull called a sulcus. Elephants employ their tusks for a variety of functions, including display, competition, defense, and obtaining food, minerals, and water. It's easy to envision why natural selection would work to favor individuals with the largest tusks—and for millennia, this has apparently been the case. Sadly, poaching elephants for their ivory tusks has become a monumental problem, which not only threatens the very existence of all elephant species but has led to a very different form of selection pressure, this one clearly due to the influence of humans.

In 2021, Princeton University professor of ecology and evolutionary biology Shane Campbell-Staton and his colleagues published a paper in *Science*. In it, they demonstrated that during the Mozambican Civil War (1977–1992), the African elephant population of Mozambique's Gorongosa National Park experienced a precipitous decline—from approximately 2,500 to around 200 individuals—as elephants were slaughtered by poachers. As a result, a strong human-driven selection for tusklessness arose, with tuskless females more than five times more likely to have survived during the period of 1978 to 2000. Elsewhere in Africa during this time period, similarly steep drops in elephant populations were taking place. Researchers found that in South Luangwa National Park (SLNP) and the Lupande Game Management Area, both in Zambia, the elephant population declined by about 12,500 individuals between 1987 and 1988. Researchers also determined that the percentage of tuskless females in those areas increased from 10.5 percent in 1969 to 38.2 percent in 1989. There were no reports of naturally occurring tuskless males, and to understand why, it's necessary to review some basic genetics.

Gender in mammals is determined by a pair of sex chromosomes (X and Y); females typically have two X chromosomes, and males one X chromosome and one Y chromosome. During the fusion of egg and sperm cells, one sex chromosome from each parent is inherited by the offspring, including an X from its mother. If that offspring receives an X from its father, it will be a female (XX). If a Y, it will be a male (XY). Since researchers found no tuskless male Mozambican elephants, Campbell-Staton proposed that this tusklessness was consistent with a mutation on the X chromosome, specifically in the genes involved in tooth formation, and that this mutation was lethal (i.e., the baby would die or be stillborn) when it was the only copy of the X chromosome (as in XY males) and nonlethal (i.e., the baby would live) when the female's second X chromosome did not carry the mutation.

The hypothesis was strengthened significantly when it was discovered that mutations to the corresponding genes in female *humans* can lead to the absence of upper lateral incisors —the same teeth that grow into elephant tusks. The mutations are also lethal in human males.

Ultimately, it appears that the incidence of tusklessness in these species will continue in areas where poaching remains a threat to elephant populations. As large-tusked females (who formerly held the highest status in breeding herds) are killed for their trophy-sized tusks, lower-status, tuskless females will be left to survive to reproductive age, mate, and pass on their genes.

The same lethal mutation responsible for the absence of tuskless male elephants in Mozambique also seems to be widespread in other African locales. According to Campbell-Staton and his colleagues, "Tuskless males do not occur in Gorongosa or in surveys of large sample sizes from Africa's most intensively studied elephant populations."

As for Asian elephants, human-related stresses are similar, but they've reportedly been around a lot longer than those imposed on African elephants in modern times. This may explain why most female Asian elephants have small tusks, which extend only an inch or two in length. Known as tushes, these tiny tusks are brittle and break easily. Compared to their African counterparts, few male Asian elephants have large tusks, and many are tuskless. The percentage of tuskless males in a population can vary significantly with locality, possibly reflecting regional variation in past ivory-hunting practices.

All of this information led South Africa–based medical researcher E. J. Raubenheimer to hypothesize that three thousand years of humans interacting with Asian elephants—either hunting them for their ivory or domesticating them—led to a similar, though slightly different, human-driven process than what has been experienced more recently in Africa.

He proposed that in the Asian species, an as-of-yet-unidentified mutation (or mutations) in the presence of serious selection pressure from humans (in this case, killing off individuals with large tusks) resulted in male *and* female Asian elephants with either small tusks or no tusks at all.

[5]

Fangs a Lot

Thou call'st me dog before thou hadst a cause,
But since I am a dog, beware my fangs.
—WILLIAM SHAKESPEARE, *THE MERCHANT OF VENICE*

ON SEPTEMBER 25, 1957, renowned herpetologist Karl Patterson Schmidt, a longtime curator at Chicago's Field Museum, was asked to identify a live snake that had been recently delivered to the museum. The specimen had been sent over by the director of the Lincoln Park Zoo, Marlin Perkins, who would later go on to considerable fame as the host of *Mutual of Omaha's Wild Kingdom*. Perkins knew that the snake had come from Africa, and his feeling was that it was likely a juvenile specimen of boomslang (*Dispholidus typus*).

"I took it from Dr. Robert Inger without thinking of any precaution," Schmidt wrote in his journal that day, "and it promptly bit me on the fleshy lateral aspect of the first joint of the left thumb. The mouth was widely opened and the bite was made with the rear fangs only, only the right fang entering to its full length of about 3 mm." Expressing doubt that the snake could generate enough venom to harm him, Schmidt sucked on the wound vigorously but sought no further medical attention. He did, however, continue to document the experience in his journal.

•••

FANGS ARE LONG, sharply tipped teeth. When they occur in mammals, fangs are generally elongated canine teeth, projecting downward but with their tips curving backward. This recurved condition may have evolved to take advantage of the bite victim's initial, and often reflexive, reaction: to pull away from the mouth that bit you. By doing so, the fangs are driven deeper into the flesh of the victim, whose further struggle either sets these canines deeper still or causes severe tissue damage as what were initially puncture wounds enlarge into a tear.

In addition to puncturing and tearing, fangs are also employed in threat displays (*snarl!*) and fighting.

In nonmammals, like snakes, fangs are hollow or grooved structures, commonly used to deliver venom to a victim, hypodermic-style. At the mention of fangs, many readers will immediately think about vipers, like rattlesnakes. But just as not all snakes are venomous, not all venomous snakes are vipers.*

OPISTHOGLYPHOUS SNAKES

The boomslang that bit Karl Schmidt was an opisthoglyphous snake, one of two groups that contain venomous snakes that are significantly different from the vipers. Commonly known as rear-fanged snakes, opisthoglyphs, like boomslangs and twig snakes (*Thelotornis* spp.) are characterized by having two pairs of grooved, venom-conducting fangs located toward the back of their mouth. (See figure 12A.) Although

* And just as vipers are not the only venomous snakes, snakes aren't the only creatures with fang-delivered venom. In spiders, for example, fangs aren't teeth; they're modified jawlike structures called chelicerae. And although black widow spiders (*Latrodectus mactans*) and their relatives have gained most of the notoriety, with their neurotoxic bites (i.e., nervous-system-damaging), around thirty species of spiders have been responsible for human deaths. Anyone who might now be considering a spider-stomping spree should also be aware that there are over forty-three thousand species of spiders worldwide, which means that only around one out of every fourteen hundred species of spiders can potentially kill you. And by way of a helpful reminder, many spiders feed on harmful insects like flies and mosquitoes. So give these critters a break—they're on our side!

this arrangement works quite well for dispatching smaller prey, which can be easily manipulated into position so that they contact the fangs, things can get a bit dicey for opisthoglyphs attempting to envenomate larger-bodied prey.

Schmidt's boomslang was a native of sub-Saharan Africa and a member of the superfamily of snakes known as Colubridae (more than twenty-five hundred species, and most are harmless). Timid by nature, boomslangs spend most of their lives in trees, making their alternative common name, the green tree snake, an appropriate one. As such, they are also arboreal hunters, feeding on lizards, chameleons, birds, and small mammals.

Schmidt knew that rear-fanged snakes must literally chew on their victims to work the venom in deeply. And while there had been human deaths attributed to at least three species of opisthoglyphs, the snakes' fang placement, generally small body size, and petite venom glands led most mid-twentieth-century herpetologists to consider rear-fanged snakes relatively harmless to humans.

Schmidt also knew that a potential problem with rear-fanged snakes was that there were some big ones—including boomslangs—though not the individual that had bitten him. With relatively large fangs and a wide gape (measuring up to 170 degrees), a six-foot-long boomslang can quickly deposit a large volume of incredibly potent hemotoxin into a vertebrate circulatory system—including one being hauled around by a bulky, bipedal bite recipient. Through a process called disseminated intravascular coagulation (DIC), abnormal blood clots begin to form in blood vessels throughout the body of the victim, blocking the smallest vessels. There are so many clots that the body uses up all of its normal reserve of clotting factor—the long list of chemicals that must come together in just the right order to produce a normal blood clot.

This complex clotting process, which falls under the heading of

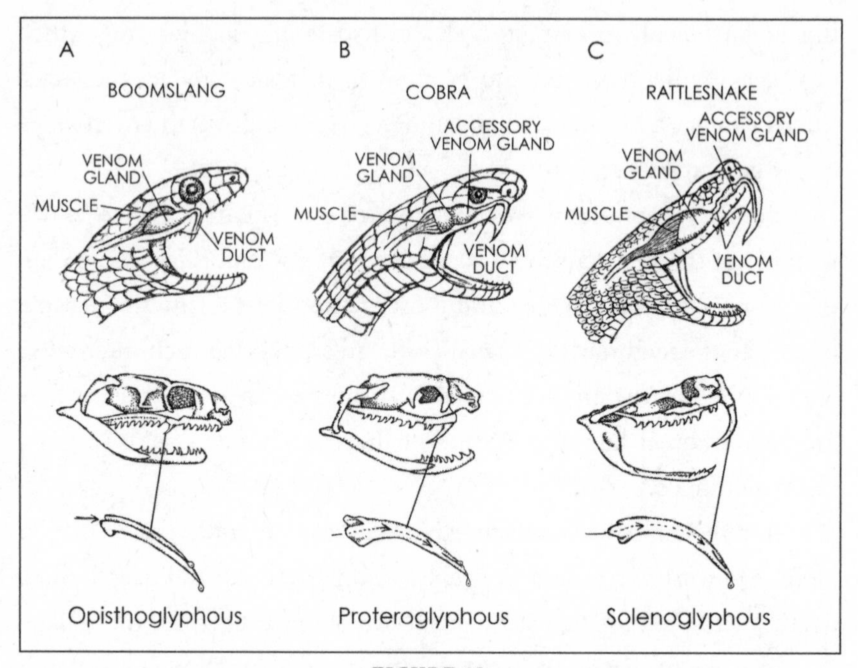

FIGURE 12

"Chemical Cascade," has been memorized by countless human anatomy and physiology students. Presumably, the stepwise nature of the process evolved as a safeguard against blood clots forming accidentally. Here, though, in an organism envenomated by an opisthoglyph, the body's normal clotting mechanisms are overwhelmed by the hemotoxin. Usually, the body in question belongs to a bird or a lizard, which dies in only a few minutes. But having entered the bloodstream of a human, with a body hundreds of times larger, only after what can be several disarmingly long hours of seeming ineffectiveness, do the venom's catastrophic effects on the circulatory system begin to take place.

By the late afternoon of September 25, 1957, Karl Schmidt's notes had pretty much transitioned into a list of symptoms that he had begun to experience. Returning home by train, he recorded "strong nausea but without vomiting." An hour later, he described "strong chill and shaking followed by fever of 101.7" and "bleeding of mucus membranes in the mouth." Schmidt spent a difficult night ("violent nausea and vomiting," "urination at 12:20 AM mostly blood") but reported feeling better in the morning, even calling the museum to assure colleagues that he'd be coming in later that day. At six thirty, he wrote, "Ate cereal and poached eggs on toast and apple sauce and coffee for breakfast. No urine with an ounce or so of blood about every three hours. Mouth and nose continuing to bleed, not excessively."

Schmidt's recording of the event stopped there. Shortly after lunch, his condition worsened dramatically. He began struggling to breathe and soon lapsed into a coma. Resuscitation attempts were futile. At just before three o'clock on September 26, Karl Schmidt was reportedly dead upon arrival at the hospital. An autopsy revealed that he had suffered massive internal bleeding from nearly every major organ.

Some readers might question why an expert herpetologist like Schmidt did not seek immediate treatment for a bite from what is now known to be one of the world's deadliest snakes. The answer is not as simple as one might imagine. First, although the potential lethality from boomslang envenomations had been well-documented by 1957, Schmidt and other herpetologists of the time mistakenly believed that so-called rear-fanged snakes like the boomslang could not deliver a fatal dose of venom to humans.

They likely pointed to the fact that the venom glands of these snakes are generally small, and so the total volume of venom would be slight. Adding to the apparent doubt about the danger Schmidt was in was his own observation that the bite had been made by only one fang.

Schmidt also apparently had doubts that the snake Marlin Perkins had sent him was indeed a boomslang. He wrote in his journal that while the specimen's head shape and bright coloration matched the description of a boomslang, the "anal plate was undivided." Also known as the anal scale, this structure is located just in front of, and covering, the snake's cloacal opening,* and its appearance is used (or at least it *was* used) as a quick indicator to determine whether a snake is venomous ("divided" or paired anal plate) or harmless ("undivided" or single plate). Since boomslangs are indeed venomous, Schmidt was likely looking, unsuccessfully, it seems, for a divided anal plate, and he made note of that absence in his journal.

Finally, once he had been bitten, Schmidt likely believed that the antivenin specifically designed to neutralize the toxic components in boomslang venom could only be found in Africa, so that even if he were to receive immediate treatment, there would be no guarantee that he would survive.

Perhaps, then, as some have suggested, Karl Schmidt used his final hours on Earth to observe and meticulously record a truly unique experiment—one in which, after becoming the unfortunate subject, he dedicated himself to getting right—even if it was the last thing he ever did.

In an interesting coda to this strange and tragic tale, in 2017 European researchers determined that an antivenin designed for use against rattlesnakes—and available when Schmidt received his fatal bite—would likely have some degree of effectiveness against boomslang venom. But since physicians back then knew nothing of this partial cross-reactivity, it is unlikely they would have used it during efforts to save Schmidt's life. The rattlesnake antivenin had, after all, been developed to treat victims envenomated by vipers living a half a world away from this particular rear-fanged African import.

---

* The cloaca is a single, common opening for the digestive, urinary, and reproductive tracts. It is commonly found in nonmammal vertebrates as well as egg-laying monotremes, like the duck-billed platypus.

COBRAS AND THEIR KIN

Among the so-called front-fanged snakes are proteroglyphs, like cobras, mambas, coral snakes, and sea snakes, which have smallish fangs fixed in place at the front of a pair of upper jawbones, the right and left maxillaries. (See figure 12B.) Snakes with this jaw anatomy belong to the family Elapidae, which is composed of approximately 360 species—all venomous. Many elapids are either famous or infamous (depending on where you happen to be standing) for their distinctive threat displays, during which they rear upward and spread a pair of flattened neck flaps. Like their rear-fanged opisthoglyph cousins, after a bite proteroglyphs generally hang on to their prey until the venom has taken effect. Unlike the smaller (but apparently just large enough) venom glands of rear-fanged snakes, the venom-producing glands of a cobra or a mamba have a large central cavity (called a lumen) that can hold a substantial volume of ready-to-inject venom.

The ingredients in a venom cocktail can vary tremendously, depending on which of the estimated one hundred thousand plus venomous animal species you're referring to. There are, however, some basic components. These include proteins (*especially* enzymes), chains of proteins (called peptides), and nonproteins (like carbohydrates, lipids, and metal ions). That said, a 2019 study put our current knowledge of venoms into perspective, stating that "most known toxins have been described only incompletely." As for their pathophysiological (i.e., nasty) effects, the list includes: neurotoxicity (which can occur at several sites along the path leading to the transmission of a nerve impulse), hemotoxicity (which includes both pro- and anticoagulant activities of the blood), cytotoxicity (cell toxicity), and myotoxicity (muscle toxicity). There are different delivery systems that have evolved to administer venom—from the glands where it is produced and stored through tubelike ducts leading to fangs of various form.*

By now, some readers may have also wondered about the difference between venom (like the substance that took the life of Karl Schmidt) and poison (hemlock tea, anyone?). Venom needs to be introduced into the circulatory system to have its desired (or undesired) effect. Teeth and venomous spines (the needlelike anatomical structures wielded by some fish) are the means by which venom commonly gets into the bloodstream. Poison, like the hemlock concoction Socrates was sentenced to drink in 399 BCE, can enter the body by being ingested, inhaled, or absorbed through the skin. As for venomous vs. poisonous animals, venom is administered aggressively, while a poison arrow frog, in contrast, has no control over the fact that consuming it can poison the consumer. With those differences in mind, it's clear that snakes administer venom, not poison. Venom functions range from the capturing, killing, and digestion of prey to defense against predators—including those who choose to antagonize the wrong critter.

Among the proteroglyphs, several species of cobras have a cool fang-related adaptation: they can defensively spit venom in a pressurized horizontal stream, often into the eyes of a potential predator or harasser. In these snakes, a secondary function of their venom is to cause pain. The evolutionary tweaks that allowed for this impressive spitting behavior relate to modifications of several jaw muscles as well as a rounder shape of the venom discharge orifice located near the tip of the fang.

The king cobra (*Ophiophagus hannah*) is not only the largest venomous snake, commonly attaining lengths of ten to thirteen feet, but

---

* Exocrine glands, like sweat glands, produce substances (like sweat) that are transported to their destinations via tubes called ducts. Alternately, endocrine glands (a.k.a. ductless glands) dump their products (like the chemical messengers known as hormones) directly into the blood. There, these endocrine secretions are transported via the circulatory system to sites where they pass on their messages (like "Grow" or "Beat faster") to cells whose surface receptors allow them to read the messages.

it also holds the record for venom volume delivered. A single bite from this native of South and Southeast Asia, can deliver around 7 milliliters (1.5 teaspoons) of neurotoxin into a victim. Different from the hemotoxic venom that killed Karl Schmidt, cobra venom acts by blocking nerve impulses traveling to and from structures like muscles. The resulting muscular paralysis from a cobra or mamba bite can turn deadly if, for example, the venom blocks nerve transmission from the phrenic nerve to the dome-shaped, muscular diaphragm—the most important of the breathing-related muscles in mammals.

Though bites from elapids cause thousands of deaths each year, from a pharmacological perspective there is a positive side to the neurotoxicity associated with their particular form of venom. Medical researchers who have analyzed elapid venom have determined that these substances and their mode of action could be used to produce nerve-blocking anesthetics to treat conditions like arthritis—a generalized term for inflammation of the joints.

Here, the anesthetic effect takes place through the interruption of nerve impulses that begin with the activation of relatively unspecialized neurons called nociceptors (better known as pain receptors). These are found in widespread locations throughout the body, in the skin, muscles, joints, and other organs. Pain receptors are stimulated by and respond to chemicals released by damaged or inflamed tissue, the brain, or white blood cells that have migrated to the problem area.

Without interruption, nerve impulses generated by nociceptors, like those located within arthritic joints, are passed along to sensory neurons. Their job is to transmit their own nerve impulses toward the central nervous system.* Eventually, these signals reach a region of the brain

* Nociceptors (*noci* = "hurt" in Latin) are sensory neurons, since they send information (about damage or potential damage to the body) *toward* the central nervous system (CNS) i.e., the spinal cord, and the brain. Alternatively, motor neurons send signals *away* from the CNS, most often to muscles, which contract in response to that stimulus. Completing the circuit, interneurons (or association neurons) are those that link sensory neurons to motor neurons *within* the CNS.

called the prefrontal cortex, which produces the perception of pain by the sufferer. Blocking the nerve transmission prevents that perception from happening (at least temporarily), since, in arthritis for example, signals that began in the joints never get to the brain. To follow this to its end, as the chemical compounds resulting in the "nerve block" are broken down by the body, the anesthetic or pain-killing effects wear off, and the signals once again travel from the arthritic joints to the brain.

None of these developments related to anesthesia are likely to be a comfort to anyone bitten by a cobra, a death adder, or a coral snake. But besides having evolved the most efficient delivery system for their venom, the last group of venomous snakes, the vipers, have also played a key role in the development of several important venom-based medications.

VIPERS

There are approximately 330 species in the Viperidae family (otherwise known as solenoglyphous snakes). Although they do not make up the largest venomous snake family, vipers are the most widely distributed. Seen throughout Africa, Asia, the Americas, and Europe, the viperids include such "party's over" favorites as rattlesnakes, copperheads, and water moccasins. Characterized by the presence of large, roughly triangular heads, vipers vary in body length: from seven to ten inches for the Namaqua dwarf adder* (*Bitis schneideri*) to around ten feet for the bushmaster (*Lachesis muta*). Within those wedge-shaped heads, the paired maxillary bones of the upper jaw are reduced to mere nubs. (See figure 12C.) But unlike other bitey snakes, each of the two maxillary bones in a viper's skull bears a single tooth of majorly impressive length (though admittedly, it helps to be a snake lover or herpetologist here). As for the largest viper fangs, that record belongs to the Gaboon viper (*Bitis*

* Also known as the spotted dwarf adder and the Schneider's adder.

gabonica), of central and western Africa. A six-foot-long specimen can have fangs reaching up to *two inches* in length, with the ability to deliver enough venom to kill thirty full-grown adult humans.

So how do vipers avoid dragging their extralong fangs across the ground? The secret to avoiding the fangs-exposed-when-not-biting saber-toothed-cat look relies on the presence of hinges at the base of each fang. This enables them to be folded, tip facing backward, against the roof of the mouth. Only when the jaws open, which they can do to an amazing angle of nearly 180 degrees, do these serpentine hypodermic needles snap forward into place. Hollow, rather than grooved, viper fangs can inject relatively massive amounts of venom, produced, and stored in a pair of Duvernoy's glands (the viper version of venom glands), located between the jaw joint and the eye. During the bite, muscles associated with the glands contract—the equivalent of compressing the rubber bulb of an eyedrop applicator—sending venom on its way toward the fangs and the victim.

In 1998, researchers Kenneth Kardong and Vincent Bels used a high-speed camera to study the mechanics of rattlesnake bites, providing a valuable example of the complexity of this highly evolved form of predation and self-defense. They noted both prestrike and poststrike behavior, but their analysis focused on the strike itself. The strike phase (see figure 13), which took approximately half a second, consisted of four stages: the extension stage, the contact or bite stage, the release stage, and the retraction stage. Kardong and Bels also observed "occasional, often complex modifications of the basic pattern: extend, contact, release/retract."

A decade later, researcher David Cundall used a camcorder to record strikes from eighty-six different viper species. After analyzing the data, he found that while generalized descriptions of viper strikes, like those provided by Kardong and Bels, were helpful, the actual event sequence often varied because of the nature of the living, moving prey—and the stochastic (or randomly determined) behavior exhibited by that prey. Cundall hypothesized that behavioral adaptations, like the ability of the viper to rapidly reposition its fangs during a strike, evolved to

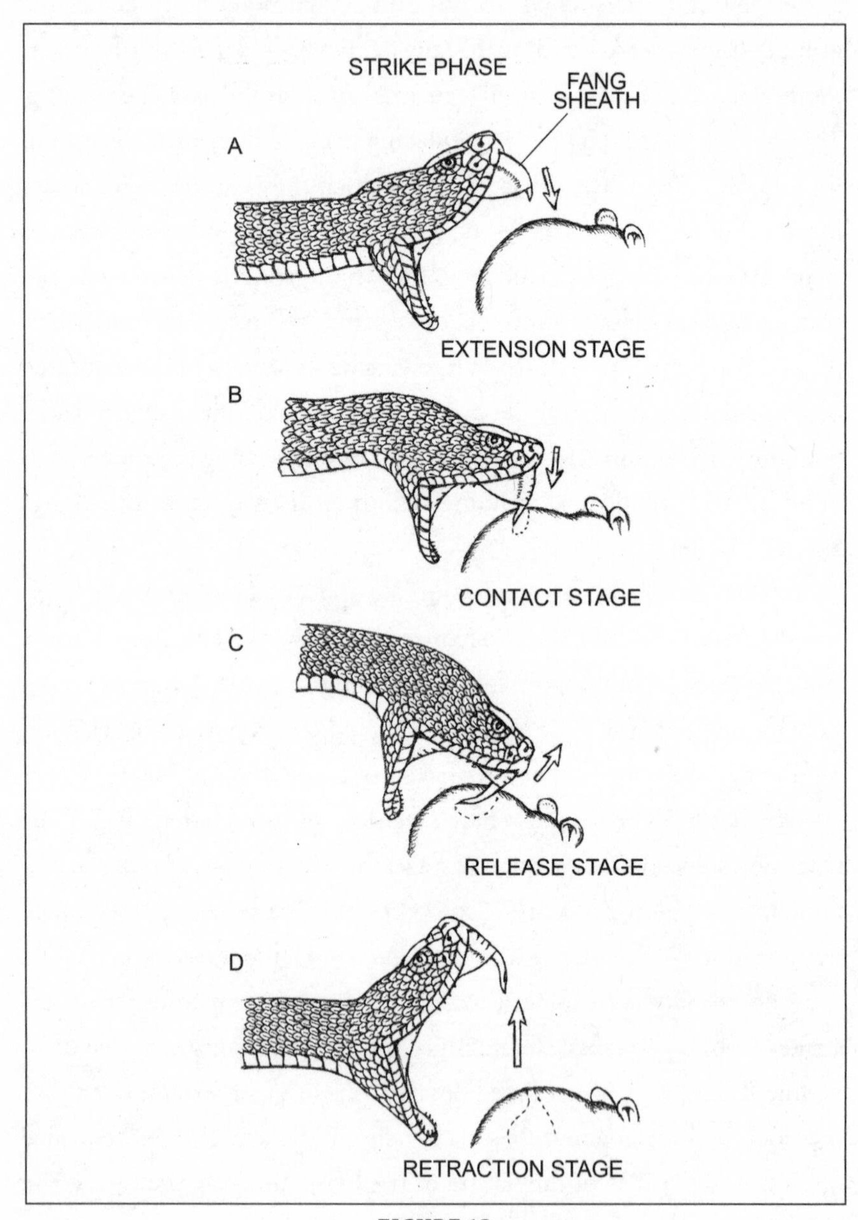

FIGURE 13

immobilize the prey as quickly as possible to reduce the potential for the prey to (1) wound the snake, and/or (2) move far away from the viper after being released.

Another fascinating aspect of viper bites and envenomation is that they are not all-or-nothing phenomena. Solenoglyphs can regulate the amount of venom they inject or don't inject, depending on their perception of the situation. Dry bites are those where there is no envenomation, due to the snake deciding not to contract the muscles typically used to compress the venom glands during a strike.

The toxic compounds in viper venom include specialized enzymes called metalloproteases.* Proteases break down proteins, and metalloproteases are proteases whose mechanism of action requires the presence of a metal, such as calcium and zinc. Metalloproteases can also have multiple toxic effects. In addition to producing systemic hemorrhaging and fouling up the blood-clotting process, they break down collagen, the most abundant protein found in vertebrates—including humans. Collagen is not only a key component of skin, but it forms a sort of internal skeleton within organs and muscles. Basically, collagen forms a connective tissue scaffold upon which other cells grow and adhere in an organized fashion.

A recent study by Harry Williams of the University of Reading and his colleagues, indicated that the problem with exposure to viper venom is often twofold, with the initial destruction of the collagen framework by metalloproteases accompanied by damage to the so-called basement membranes, upon which many cells, including myocytes (i.e., muscle cells) grow and regenerate. The result can be permanent muscle damage—requiring surgery followed by skin grafts and/or amputation.

Given the toxic effects of viper venom and other snake venom, a natural question would be: Why doesn't the toxin have an effect on the snake?

---

* The suffix "-ase" at the end of a chemical compound means that the substance is an enzyme, a type of protein that catalyzes (speeds up) chemical reactions. The substrate (what the enzyme acts on) is designated by the initial section of the enzyme's name (as in "protease").

Why, for example, don't the enzymatic, tissue-destroying properties of viper venom wreak havoc on the venom glands of these creatures?

Once again, the answers can vary depending on the species. In some vipers and elapids, the accessory venom glands produce chemical inhibitors that control the venom toxicity while that venom is being stored. In other instances, within the main venom gland itself, acid-secreting cells like the parietal cells found in the stomach lining, secrete acidic compounds that keep the pH low enough (5.4) so that the venom stored in the gland remains inactive. Additional mechanisms include the presence of antibodies to venom components circulating in the blood or the presence of modified receptors on the membrane surface of the snake's muscle cells.

Cell membrane receptors can be thought of as tiny locks that will only open in the presence of a specific chemical key. In the case of skeletal muscle cells, that key is the neurotransmitter acetylcholine, which acts as a chemical messenger, signaling the muscle cell to contract. Once unlocked (i.e., activated), cell membrane receptors are commonly involved in turning on a specific function (such as muscle contraction) or preventing that function when the key is not present.

In some venomous snakes, the shape of the muscle cell membrane receptors is tweaked, which prevents molecules in the venom (alpha-neurotoxin) from binding to the receptors instead of acetylcholine. This allows muscle cells to contract normally—something that's prevented when alpha-neurotoxin is substituted for acetylcholine. Phew!

The protective mechanisms of venomous snakes are often redundant, offering them multiple ways to avoid being harmed by their own bioweapons.

Finally, although humans can be harmed by viper venom, it also has been shown to have tremendous medicinal value. For example, the jararacussu (*Bothrops jararacussu*) is a South American pit viper that can reach lengths of over seven feet. Much feared for its venomous and often deadly bite, the venom of *B. jararacussu* is a potent cocktail of cytotoxins, hemotoxins, and myotoxins. These compounds can wreak havoc with cells, the circulatory system, and muscles, respectively, leading to amputations and failure of the renal, respiratory, and circulatory systems. But we now know that there are also substances in *B. jararacussu* venom that show promise as antibacterial agents, and other studies have shown that jararacussu venom inhibited coronavirus production in monkey cells.

Most notably, the hypotensive (i.e., blood-pressure-lowering) ability of a substance isolated from the toxin of *B. jararacussu* led to the development of the drug captopril. Approved by the US Food and Drug Administration in 1981, captopril remains one of the most important medications currently used for the treatment of hypertension. Perhaps triggered by the approval of captopril, snake venoms have become a well-recognized and important resource in the development of new drugs.

[6]

Poop on the Beach: The Good Kind

Global climate change has become entangled with the problem of invasive species. A warmer climate could allow some invaders to spread farther, while causing native organisms to go extinct in their traditional habitats and making room for invaders.

—Richard Preston

Fans of *Jeopardy* might respond to the answer "This group of vertebrates contains more venomous species than any other" with the question "What are snakes?" But they'd be wrong.

That's because while there are around six hundred species of venomous snakes, there are more than four times as many venomous fish—roughly twenty-five hundred species.* The vast majority of these creatures envenomate their prey, potential predators, and those who mishandle or disturb them, with a variety of pointy protrusions. These include tail spikes (stingrays), spines (catfish), modified fins (scorpion fish), and teeth (fang blennies). Some species, like catfish and bullheads, can inflict nasty wounds, but venom-wise, they're relatively harmless to humans. Others are equipped with venom that can cause serious medical issues and even death for those on the receiving end.

Unlucky swimmers in the Caribbean and along the eastern coast of

---

* Percentagewise, venomous snakes come out on top, since there are only about thirty-nine hundred snake species, with six hundred venomous, around 15 percent. There are roughly thirty-four thousand species of fish, with twenty-five hundred venomous, or around 7 percent.

the United States are literally running into a venomous spine-related problem that began in the mid-1980s, when two closely related and nearly indistinguishable species of lionfish (*Pterois volitans* and *Pterois miles*), native to Indonesia, began appearing off the coast of Miami, Florida. Their introduction likely occurred after tropical aquarium owners grew tired of having these admittedly beautiful additions to their saltwater tanks gulp down all their other fish—and threw them in the ocean. With no Atlantic predators, the newly introduced lionfish (armed with venomous dorsal, pectoral, and anal fin spines) found the warm southern waters to be a great place to hang out, feed voraciously, and reproduce. In recent years, these sit-and-wait ambush predators have become a serious invasive species, threatening dozens of important native fish like snappers and groupers, whose young are swallowed whole.

Also at peril from the lionfish onslaught are highly specialized species like parrotfish and surgeonfish and the coral reefs they graze upon. Competing with coral for space, algae can smother a reef if its growth is left unchecked. Both parrotfish and surgeonfish benefit the reef by grooming algae from its surface (see figure 14).

Parrotfish is the common name for approximately ninety-five marine species belonging to the family Scaridae. They are equipped with impressively strong beak-like structures that enables them to snip off pieces of

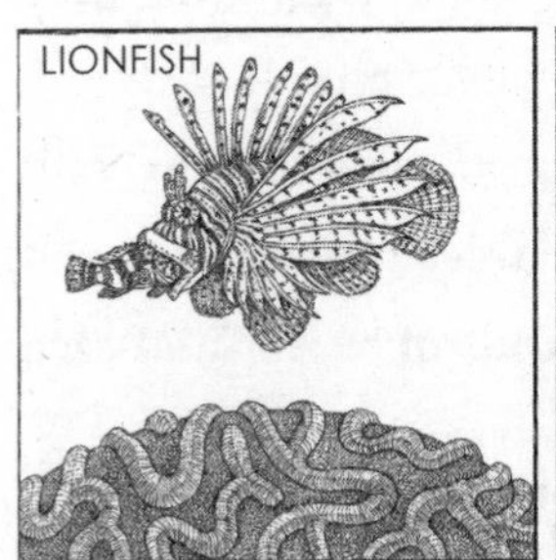

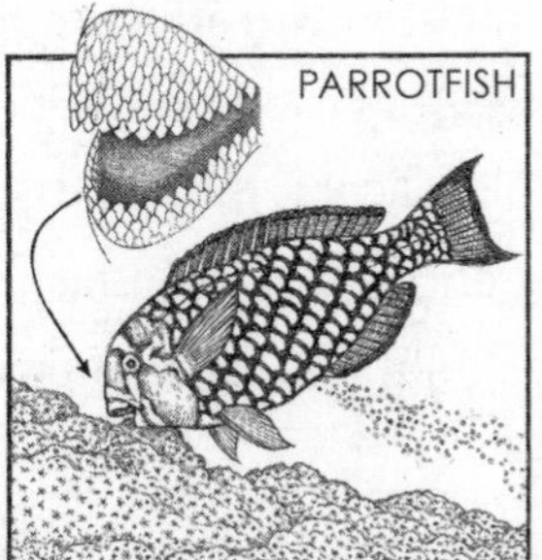

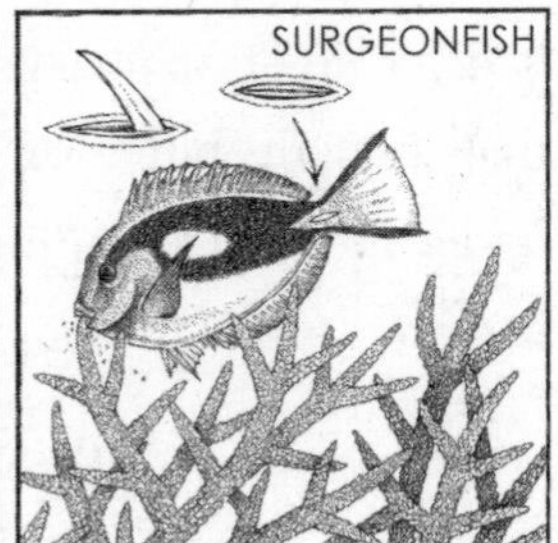

FIGURE 14

dead algae-covered coral. Recent studies by Matthew Marcus, staff scientist at Lawrence Berkeley National Laboratory, and his colleagues showed that one factor in the parrotfish's coral-crunching ability results from crystals of fluorapatite (a mineral made up of fluoride, calcium, and phosphorus) woven into microscopic bundles in its teeth. These bundles can be considered building blocks for the production of enamel, in humans, and enameloid, in species like the parrotfish. In these reef dwellers, the enameloid covers the business end of the roughly one thousand miniature teeth present in the fish's mouth. Parrotfish teeth are uniquely stacked up around fifteen rows high, with each tooth fitting into the rear of the tooth in front of it and cemented to the teeth surrounding it. To the naked eye, this arrangement forms a beak that exhibits an impressive combination of strength and durability. As the older teeth at the beak edges are worn away, they're replaced by teeth in the rows located just behind them.* As a result, the parrotfish is well adapted to bite off small chunks of algae-covered coral.

While the parrotfish reaps the nutritional benefits of consuming the algae and cyanobacteria (bacteria-like microorganisms capable of photosynthesis) growing on the coral, the reef benefits from the removal of the algae coating it, as well as the removal of sections of dead coral. This provides the reef with space for new coral growth, though it should be noted that so-called corallivorous species like parrotfish do consume live coral as well. As the chunks of coral move from the mouth rearward and into the pharynx, they enter the parrotfish's version of a grinding mill, which consists of pharyngeal jaws and rows of flattened pharyngeal teeth that crush the bitten-off bits of coral into sand—a phenomenon that falls under the broad heading of bioerosion. In fact, larger species of parrotfish are capable of pooping out hundreds of pounds of beach sand each

* If this makes parrotfish sound vaguely like shark teeth, they're not—since shark teeth do not fit into each other, nor are they connected to each other to form a sort of beak.

year. This benefits humans who want to lie on that poop or form it into miniature castles.

Surgeonfish take a gentler, though no less useful, approach to algae removal and reef maintenance than the parrotfish. Belonging to the family Acanthuridae, which consists of around eighty-five species, these pancake-shaped reef dwellers have no tooth-lined pharyngeal jaws to grind up coral bits. Instead, they're equipped with a set of more delicately built, spoon-shaped teeth, which they use to nip off filaments of algae. One might think that the name surgeonfish derives from the surgeon-like precision with which they remove algae from a reef. But it's actually a reference to a pair of scalpel-sharp venomous spines located on each side of their tail. These blades normally fit into horizontal grooves along the body, but when threatened the surgeonfish flashes them, switchblade-style, while whipping its body from side to side.

Unfortunately, invasive lionfish feed on juvenile surgeonfish and parrotfish, and other species that help curb algae growth, putting these native fish and the coral they service in danger of extinction. In one study conducted in the Bahamas, researchers showed that a single lionfish could reduce a measurement known as recruitment—the transition of fish from one stage of life to another stage—by 79 percent within five weeks. In their experiment, of the thirty-eight species of native reef dwellers observed, twenty-three of them experienced reduced recruitment. These types of numbers led James Morris Jr. of the National Oceanographic and Atmospheric Administration to conclude: "The invasion of lionfish . . . may prove to be one of the greatest threats of this century to warm temperate and tropical Atlantic reefs and associated habitats."

Currently, lionfish are performing the equivalent of a fin-propelled march across the Gulf of Mexico and the Caribbean, and up the Eastern Seaboard of the United States as far north as New York. Because invasive marine species are nearly impossible to eradicate without also eliminating native species, NOAA is promoting one innovative approach to the problem. Lionfish are reportedly quite delicious (once they've had their venomous spines *carefully* removed and have been fileted), so the agency's recommendation boils down to something akin to "If you can't beat 'em, eat 'em." As a seafood lover, I quickly followed up and located a list of online recipes that included lionfish ceviche, blackened lionfish with creamy potato salad, and lionfish with mango sweet pepper salsa.

Before transitioning from our exploration of venomous spines to venomous fangs, and by way of a brief public service announcement, I offer this warning: while contact with lionfish and their spines generally falls under the heading of "Intensely Painful Experiences," encounters with another group of venomous fish can be fatal to humans.

The stonefish (*Synanceia* spp.) holds the title as the most venomous fish in the world. Its venom, which is administered by thirteen stiff, super-sharp dorsal spines, contains a substance called verrucotoxin (VTX). This potentially lethal glycoprotein* can produce cytotoxic, hemotoxic, and myotoxic effects that include, pain, seizures, convulsions, paralysis, and respiratory and cardiovascular damage. A clear take-home message for those swimmers exploring the reefs where these well-camouflaged fish are found (i.e., coastal regions of the Indo-Pacific, plus New Guinea and Australia) is to familiarize yourself with any potentially dangerous local species and be *extremely* careful where you tread and what you touch.

Readers may have already noticed the absence of teeth from the fishy arsenal used to envenomate prey and ward off predators. And to make it clear, though they are bony in composition, spines aren't teeth. In fact, only two genera of fish are believed to have a venomous bite.

---

* Glycoproteins are molecules composed of a protein with attached chains of carbohydrates (sugar). They are, for example, major components of viruses, and are important in both infection and immunity.

One is a strange little deep-sea eel, *Monognathus* (Greek for "one jaw"). This large-mouthed fish has a lower jaw, but the premaxillary and maxillary bones that usually make up the vertebrate upper jaw are absent. In larval stages, the upper jaw of *Monognathus* is present, only to be reabsorbed as the fish reaches maturity. Adults (but not larvae) are equipped with a hollow, centrally located ethmoid fang. This structure, which projects from the ethmoid bone (the anterior-most bone of the braincase), is surrounded at its base by toxin-producing glandular tissue. As with many deep-sea species, relatively little is known about the habits of *Monognathus*, including how its fang is employed during the bite—presumably to secure its unknown prey, serpent-style, while the venom (whose composition and bioactivity are also unknown) takes effect (see figure 15).

There is also much that we do not know about *Meiacanthus*, the only other example of a fish capable of inflicting venomous bites. But what is known about this pint-sized reef denizen is as fascinating as its common name—fang blenny—is appropriate.

The fang blenny's mouth is equipped with a pair of enormously enlarged fangs, which curve up from their fixed positions on the lower

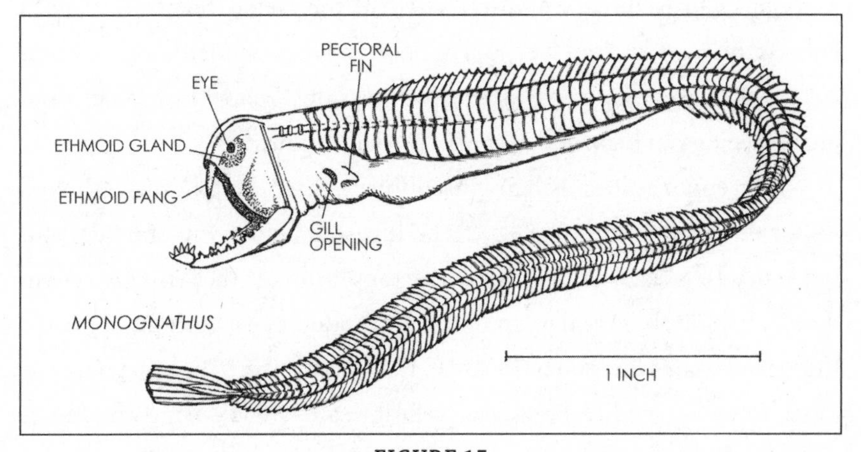

FIGURE 15

jaw. Within the fang-blenny genus *Meiacanthus*, all three species are venomous. The best-known of these is the yellowtail fang blenny (*Meiacanthus atrodorsalis*), whose peculiar behavior was first studied by zoologist George Losey during the early 1970s, on the Eniwetok Atoll in the Marshall Islands.

The yellowtail fang blenny is a stunningly beautiful and surprisingly innocent-looking fish, typically measuring a little less than five inches in length. The anterior 50 to 70 percent of its body is blue gray in color, with the remainder, including its lobed and tapering tail, colored bright yellow. According to Losey, these fish swim just above the coral, in "loose aggregations of up to 50 individuals."

Readers who might be gearing up to hear about how fang blennies use their namesake dentition to battle equally ferocious reef dwellers—slashing, puncturing, and finally scarfing them down—may be disappointed to learn that *Meiacanthus* feeds on plankton, red and brown algae, and coral mucus. Any disappointment, though, will be temporary. As it turns out, the fang blenny's fangs, and the venom delivered by them, get busy only after the little fish are swallowed whole by large piscivorous predators like groupers (*Epinephelus* spp.). (See figure 16.)

I spoke to Nicholas Casewell, chair of the Department of Tropical Disease Biology at the Liverpool School of Tropical Medicine and first author of a 2017 paper on aspects of anatomy, behavior, ecology, and pharmacology in blennies—especially *Meiacanthus*.

"The venom system is being used for defense," he told me. "Though it's an extremely risky strategy to be ingested before you start defending yourself." Casewell explained how most animals that employ venom defensively, like bees and wasps, do so by producing an amount of pain in the victim that is disproportionate to the wound size. "This is a good way to stimulate learned responses as well," he said, since during any future encounters the previously stung victim avoids the creature that stung it.

FIGURE 16

It's also the reason why many creatures that can deliver a painful bite or sting are strikingly colored and/or have distinctive color patterns that can be recognized quickly (so-called warning coloration).

Casewell told me that there were no pain responses in the tests they conducted on fang-blenny bite victims, but he said, "What we did see was that the bites stimulate a potent but quite transient drop in blood pressure." The drop in pressure is caused by two naturally occurring substances in the venom: neuropeptide Y (which acts on the vertebrate central nervous system) and proenkephalin (an endogenous opioid).*

So instead of stimulating pain receptors, fang-blenny toxin appears to target the circulatory system. Because of the small size of the molecules involved, the toxin rapidly diffuses into the blood, producing short-term

* "Endogenous" means that the substance, in this case an opioid, originated from within the organism.

effects not unlike those associated with opioids. Half a century ago, George Losey described the effect on the fang blenny's finny victims this way: "The typical reaction after taking an *M. atrodorsalis* into the mouth was violent quivering of the head with distension of the jaws and operculi. The fish frequently remained in this distended posture for several seconds until the *M. atrodorsalis* emerged from their mouth. Frequently, the *M. atrodorsalis* were little harmed by the experience."

Remembering how the blood-pressure-lowering drug captopril was produced from compounds in snake venom, I asked Casewell if anyone had investigated the possible therapeutic use of fang-blenny venom.

"Because their effects are so rapid . . . and they clear from the body extremely quickly," he said, it would be difficult to determine. He explained that once these bioactive molecules produce their effect, they are immediately broken down by organs like the liver, to be excreted or recycled.

According to Casewell, though, "*Meiacanthus* has essentially repurposed something that they already had—for use as a weapon."

This is fairly typical of what occurs in other venomous creatures; substances normally involved in other functions are either weaponized or injected at a dose that causes a far different effect from the normal physiological effects. In *Meiacanthus*, those normal functions would have related to the regulation of their own blood pressure, pain reception, and inflammation.

In yet another example of convergent evolution, the active compounds in fang-blenny venom were identified in two creatures that couldn't be less fishlike: neuropeptide Y was found in the venom of cone snails (*Conus betulinus*), and proenkephalin was found in the venom of some scorpion species. "There are certain protein or peptide types that come up time and time again," Casewell said, "being recruited or utilized or weaponized in the venoms of completely different animals."

I asked Casewell if, like venomous snakes, fang blennies had systems that prevented them from being harmed by their own venom. He explained that while venomous snakes had been well studied (due to their medical importance), fang-blenny venom glands and those of many other venomous animals had not. There have been some micro-CT scans and microscopy, but there is much more to learn.

"Interesting grad student project," I commented.

"Absolutely!" Casewell responded.

What followed was a mutual lament about how the explosion of molecular-based research had made funding more classical studies in morphology and ecology *extremely* difficult to obtain. Casewell said there might be serious interest in the biochemistry of a certain venom, for example, but without studies in which this information is examined in the context of the natural environment (i.e., where the venom is used) there can never be a full understanding of what researchers are seeing in the lab. As a result, researchers often must resort to speculation and inference.

When we'd finished complaining, Casewell told me about one aspect of fang-blenny tooth-related biology that *has* been studied. Although all fang blennies have enlarged teeth, only the genus *Meiacanthus* has a venom system associated with it. Casewell said this indicates to him that "the fangs did not evolve for delivering a venomous bite," nor do they appear to serve a current role in foraging. His hypothesis is that large, ungrooved fangs evolved as a type of mechanical protection—to inflict a painful bite on any predators that might consume them—no venom needed. Only later did chemicals that were already serving other functions in the fang blenny's body evolve new roles as chemical weapons.

But from no fangs to fangs to venomous fangs (in one genus), the evolution of fang blennies has one more intriguing stage to examine, and that is mimicry—in fact, fang blennies display two of the three known forms of mimicry. Some species are Batesian mimics, named for the British naturalist Henry Bates (1825–1892). During his exploration of South America, Bates determined that some of the creatures he had brought back resembled species that were toxic or venomous. This, Nick Casewell told me, means "I'll leave you alone because I think you look dangerous."

There are many examples of Batesian mimicry throughout nature. Perhaps the most famous is how the nonvenomous scarlet king snake, *Lampropeltis elapsoides* (the mimic) resembles the venomous eastern coral snake, *Micrurus fulvius* (the model). Remember that warning coloration in wasps and bees? Well, there are harmless moths and flies that mimic the color and behavior of those venomous fliers, thus helping themselves to avoid being eaten.

Within the fang blennies, there are different Batesian mimics for each of the three species of *Meiacanthus*. In each case, the distinctive coloration of the nonvenomous mimic (which resembles the venomous model) warns larger fish not to eat them—or else.

Another form of mimicry found in nature, but not in fang blennies, is Müllerian mimicry, proposed by German naturalist Fritz Müller (1822–1897), in which two or more noxious or dangerous animals have evolved to mimic each other's very real warning systems. My personal favorites are the monarch and viceroy butterflies. Until 1991, it was believed that the viceroys (*Limenitis archippus*) were Batesian mimics of the foul-tasting monarchs (*Danaus plexippus*) but in a paper in *Nature*, researchers David Ritland and Lincoln Brower showed that viceroys are as unpalatable as monarchs, a finding that transitioned both butterflies into the realm of Müllerian co-mimicry.

Though Batesian and Müllerian mimicry are quite common, a third form of mimicry, and one found in fang blennies, is quite rare. That phenomenon is known as aggressive mimicry, in which predators or parasites mimic a model to fool their prey. An example of aggressive mimicry occurs in two nonvenomous species of fang blennies that resemble *Meiacanthus*. Occasionally, one of these fish will swim near a larger reef fish that happens to be hanging out. Of course, the big guy won't eat the mimic for fear of the consequences—that's an example of Batesian mimicry. But as the tiny fish sidles in closer and closer, things suddenly take a turn for the strange, when the *Meiacanthus* mimic darts in and bites off a scale or a tiny section of the larger fish's gills, which it consumes. Casewell summed up this example of aggressive mimicry as "I will exploit the fact that I'm being left alone, for my own benefit."

Finally, as mentioned earlier, only one other fish, the deep-sea eel *Monognathus*, has a venomous bite. These fish are in no way closely related to fang blennies, but they do share one important characteristic: a dorsal fin that has been reduced in size and lacks the spines commonly used for defensive purposes. One possibility is that with limited options for structures that might someday become spine-related venom systems for *Monognathus* and also *Meiacanthus*, the presence of large fangs may have provided both with a different path—a path that resulted in the evolution of venomous bites.

[7]

Shrews: Tiny in Size, Major in Attitude

They don't leave much, do they?

—THORNE SHERMAN IN *THE KILLER SHREWS*

WHEN I WAS a graduate student in zoology researching bats, it was no surprise that Mammalogy became my favorite required course. Taught by Deedra McClearn, the wildly entertaining and equally brilliant functional ecologist, she was assisted by a pair of superb teaching assistants: Karen Reiss, well on her way to becoming an expert on toothless mammals, and Greg Graffin, an evolutionary biologist who studied fossil fish. Greg also happened to be the founder, songwriter, and lead vocalist of the internationally known punk band Bad Religion.

But beyond the eclectic faculty and staff, a key reason I found the course to be so engaging was the fact that we made frequent visits to the Arnot Teaching and Research Forest, located some fifteen miles south of Ithaca, in Central New York. The forest consists of more than four thousand mostly hilly acres of mixed hardwoods, over which are scattered ten ponds and a creek. During each visit, we students would break up into small teams, laying out dozens of live traps, located within one or more of the thirty-seven predetermined grids that divided the forest into "lots."

During my first visit, I was admittedly dismayed to learn that after painstakingly setting out our traps, we would have to check them again four hours later. If we didn't check them, McClearn explained, the species we would be capturing most frequently would starve to death. That species was *Blarina brevicauda*, otherwise known as the northern short-tailed shrew.

Resembling a stout field mouse, *B. brevicauda* is a tiny burrowing mammal with inconspicuous ears and minuscule eyes well hidden behind a long narrow snout. Though rodent-like in appearance, shrews had been long classified as insectivores—a grouping that now reflects dietary preference but *not* evolutionary relatedness or ancestry. As a so-called nontaxonomic group, the order formerly known as Insectivora (which contained shrews, moles, and hedgehogs) has been largely abandoned by modern taxonomists.*

Weighing in at a little over an ounce, and with a body length of three to four inches, *B. brevicauda* is one of the largest of the roughly four hundred species of shrews in the family Soricidae. The Etruscan shrew (*Suncus etruscus*), which weighs around 0.063 ounces, is not only the smallest terrestrial mammal but also holds the record for highest heart rate, at approximately fifteen hundred beats per minute (or twenty-five beats per second!). Of course, shrews bite, but three species of North American natives (genus *Blarina*) and two species of Old World water shrews (genus *Neomys*) share a characteristic that makes them *nearly* unique among mammals: their bites are venomous.

Before we get to shrew venom, though, some shrews, like those in the genus *Blarina*, have another cool tooth-related adaptation. They have iron in their enamel—iron hydroxide, to be specific. This can be best observed on the upper and lower occlusal surfaces of their teeth. Like an old iron fence, the crushing and grinding surfaces appear rusty

---

* This is reminiscent of the now-defunct order Reptilia, which formerly contained the snakes, lizards, crocodiles, turtles, and dinosaurs. To the dismay of nonscientist types, reptiles and even dinosaurs are no longer considered valid groupings by modern taxonomists.

red in the presence of oxygen. And, yes, this is also the reason that the iron-containing pigment hemoglobin makes our blood look red as it transports oxygen. In shrew teeth, the presence of iron is thought to make their enamel more wear-resistant, which is especially important given their nonstop eating habits.

According to Lázaro Guevara, a research associate at Instituto de Biología de la Universidad Nacional Autónoma de México, "Shrews do not inject the venom, as we commonly see in snakes." Instead, he told me, the shrew employs a pair of long, forward-facing lower incisors. At least two venomous chemicals are produced by a shrew's submaxillary and sublingual salivary glands. As venom-containing saliva travels from the glands through ducts that release it into the shrew's mouth, the saliva flows into grooves on the lip side of its bladelike lower incisors. Coupled with the action of its curved, hook-shaped upper incisors, this permits the saliva to reach the prey at the exact moment of the bite.

As in other venomous species, shrew venom is a chemical cocktail that contains several different toxins. In *B. brevicauda*, one such substance is the peptide soricidin (so named for the family Soricidae). Notably, soricidin has the ability to block muscle contraction, and *B. brevicauda* uses its venomous bite to paralyze prey *but not kill it*. What follows is known as live hoarding, which begins as the paralyzed prey are dragged off to be stored in a special hoarding chamber. The victims, which range from earthworms to small mammals like mice, are immobilized by frequent toxic bites for up to several weeks, during which time they are gradually eaten alive.

This sort of behavior relates back to the shrew's extremely high metabolic rate, which requires that the tiny mammals feed almost continuously. Only by doing so can they obtain enough nutrients and energy to fuel their hyperactive lifestyle. Live hoarding provides a hedge against conditions when prey aren't available. It also explains why we students were instructed to check our Arnot Forest live traps so frequently.

Biologist Keith Carson has speculated as to why venomous shrews aren't affected by their own venom (which had become a pet question of mine): The relatively large molecules are unable to pass through the tissue lining the shrew's mouth and into the circulatory system. Subsequently, once the venom is ingested, the shrew's digestive enzymes and stomach acid breaks down the venom into inactive forms. Seems reasonable to me.

Given my love for zoology, horror, and Hollywood, I can't end a discussion on shrews without putting in a serious plug for my all-time-favorite cheesy horror film, *The Killer Shrews* (sorry, *Plan 9 from Outer Space* fans). Dropping off supplies at a remote island, our hero Captain Sherman (actor James Best) discovers a mad scientist experimenting on a serum to shrink humans—this in a perfectly reasonable attempt to decrease demands on the world's food supply. As these things go, he has instead produced a pack of ravenously hungry, hound-sized shrews. This size comparison is easily confirmed by the filmmaker's clever use of real hounds covered in what look to be sections of carpet and fitted with three-inch extensions on their canine teeth.

The Killer Shrews was shown as a double bill in 1959 with *The Giant Gila Monster*, an exciting pairing, given that both of the featured creatures share a predilection for venomous bites. Shy inhabitants of the American Southwest, Gila monsters (*Heloderma suspectum*) envenomate their prey—and sometimes humans who annoy them—during a powerful and lengthy bite, literally chewing their venomous saliva into the wound. The venom itself is a cocktail of enzymes and other compounds produced in modified salivary glands. It is not normally fatal to humans—with no reported fatalities since 1930—but that said, their bites (which are uncommon) can produce excruciating pain and a laundry list of potentially serious problems if not treated.

In real life, non-mad-scientist types examined *Heloderma* toxin for its potential medical benefits, identifying a hormone thought to enable Gila monsters to go for months without eating. They named the hormone exendin-4 (I'm thinking the name extendin-4 was already taken). The synthetic version of exendin-4 is now used to treat type 2 diabetes by increasing the production of insulin by the pancreas. Since the life-saving properties of Gila monster venom weren't known in 1959, the movie poster production team tasked with a *Killer Shrews/Giant Gila Monster* double feature opted instead for the slightly-less-than-scientifically-accurate "Super-Sonic Hell Creatures No Weapons Can Destroy!"

[8]

Bite This!

Give me a fifteen-foot crocodile any day over a bee.

—Bindi Irwin

As a kid, Greg Erickson loved dinosaurs. But his dream of becoming a paleontologist took a detour once he entered college, where he studied aeronautical engineering. After graduating, Erickson found himself working in construction at the University of Washington's football stadium, but about a year later he decided to apply to graduate schools. In a 2009 interview with *Discover* magazine, he suggested why, after he was accepted to the Department of Integrative Biology at UC Berkeley, renowned paleontologist Jack Horner took him in. "I think he liked that I knew how to use a shovel," Erickson said.

Ultimately, though, Erickson had far more to offer the field than his ability to unearth ancient sediments and the fossils they contained. He is currently a professor of anatomy and vertebrate paleobiology at Florida State University, with a range of research interests that include growth patterns in dinosaurs, how animals flourish in extreme conditions, and how fish known as mudskippers evolved terrestrial habits. But while these are certainly intriguing topics, my interest in reaching out to Erickson related to his studies on alligators, crocodiles, and their relatives. More specifically, it was his research into the biting behavior of these creatures that drew me in.

"I always tell my students to stay away from the pointy ends," Erickson told me. "Both of them."

That would have been an impossible task for Erickson, though, given that he and his colleagues had been measuring the bite forces generated by twenty-three of the most powerfully jawed species on the planet.

When I asked him how they measured force from such a famously uncooperative group, he described an instrument that he and some engineers had designed. Their "bite bar" was a rod-shaped stainless-steel device outfitted with force transducers—small sensors that can convert the forces applied to them into measurable electrical signals. This is similar to how a bathroom scale, when you step on it, converts that force into units of weight.

In the 1990s, my grad school colleagues and I attached force transducers beneath a roughly six-inch-square metal floor plate we had situated in the center of a slightly larger frame. After placing individuals from our colony of captive vampire bats onto the plate, the transducers detected the forces applied to the plate when the bat walked—or, in the case of the common vampire bat, catapulted itself into flight.

Erickson and his colleagues had done something similar with their bite bars, but here the transducers measured the forces generated by the jaw muscles of their test subjects as they bit the bar. The researchers built their instruments in a variety of sizes to accommodate the jaws of their variously sized crocodilians. Anticipating the foul mood of their test participants, the researchers attached the leather-covered bite bars to the end of a ten-foot-long PVC pole.

Given the convenient fact that he was literally touching these animals with a ten-foot pole, I asked Erickson how he felt about working with what was arguably the world's most dangerous lineup of research subjects. He told me he had grown up around animals and had learned that the key, whether you were a rodeo rider or a veterinarian, was to know the individual. "Unlike people," he said, "animals tend to be very predictable."

The larger animals were lassoed, hauled out of the water, then securely strapped to a board. After this, the researchers broke out their bite meters. "At this point, they're more than happy to bite you," Erickson added, and I assumed he was referring to the crocs. "We place the bar on their teeth, and they just explode."

"Then what?" I asked.

"Then it's just a matter of holding on so we can get our readings."

Besides concern for himself and his colleagues, Erickson was constantly thinking about the safety of the test animals. "I don't want them to get hurt either."

I nodded in agreement, having always made it a point to handle captive animals with great care. After decades of practicing the techniques involved, I eventually found myself instructing others (mostly students) on how to do so. I became a stickler for always wearing gloves when handling bats—even the smallest wimpy-toothed species—even though I had been vaccinated for both rabies and tetanus.* I regularly nag research colleagues who post online photos showing bats being handled without gloves, which sends a bad message to the public.

Just as I had done while carefully removing captured vampire bats from a net as they flashed their razor-sharp dentition, Erickson and his fellow researchers took special care to prevent their equally toothy but significantly larger subjects from thrashing around as data were collected.

"There are no 'pretty good' crocodilian researchers. In my business.

* Bat researchers commonly have their rabies vaccine titer checked, which determines whether the level of antibodies to the rabies virus circulating in their blood is high enough to provide immunity. If the level is too low, a booster shot is administered, usually every few years.

If you're not very good, you're missing an arm or something," Erickson quipped.

He also emphasized that it takes a team of experienced people to secure the larger specimens. "We always have four of five people who really know their stuff."

Once strapped down, a tap on the snout usually causes the test subject to open its mouth, after which the bite bar is placed onto the rear teeth. That's because the laws of physics dictate that a measurement taken closest to the jaw joint will provide the greatest bite force.* This placement also stimulates something akin to the knee-jerk reaction you might have experienced while being examined by a rubber-mallet-wielding physician. Here, though, the reflexive response by the crocodilian is to chomp down on the bite bar with full force. This all-or-nothing response will become important as we explore similar attempts at measuring bite forces in mammals.

Although Erickson's first experiments centered on alligators (which makes perfect sense, given his Florida location), eventually he and his coworkers were able to obtain bite-force data on all twenty-three species of extant crocodilians, a group that includes crocodiles, alligators, caimans (alligator relatives from Central and South America), and gharials (narrow-snouted piscivores).† "We try to do three to five [individuals] of each species," Erickson said, emphasizing that this practice increased the chances of obtaining a true indication of the bite forces for that species.

Ultimately, the largest animals they measured were several seventeen-foot saltwater crocodiles (*Crocodylus porosus*), which, along

---

* Conversely, the farther away from the jaw joint (or in physics, the fulcrum) and the closer to the tip of the animal's snout, the less would be the measured forces. The trade-off here is that as force decreases, speed increases, which would be a plus for long-snouted predators like gharials feeding on fast-moving prey like fish.

† There are now thought to be twenty-four species of extant crocodilians.

with the Nile crocodile (*Crocodylus niloticus*), happen to be the species involved in the greatest number of fatal and nonfatal attacks on humans.*

Erickson and his fellow researchers were interested in determining how bite forces varied between species. Before their study, there had been several hypotheses predicting that bite forces in crocodilians would vary depending on factors like tooth shape, or the length or width of the jaw—variables that were readily observable traits commonly used to identify the species in question (see figure 17).

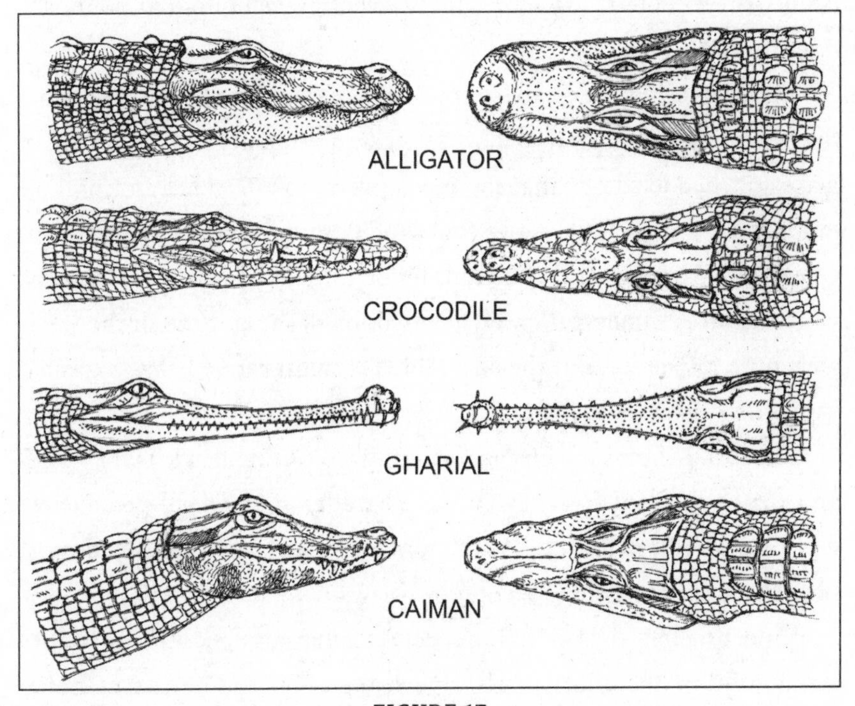

FIGURE 17

---

* Accurate numbers of crocodile-related deaths (likely in the thousands) are difficult to obtain, since many of them occur in remote or impoverished areas where they are not reported. Because it lives near large populations of humans, the Nile crocodile is thought to kill more people than its saltwater relative. Most crocodile-related deaths occur in Australia, Africa, and Southeast Asia. There have been twenty-three deaths from American alligators (*Alligator mississippiensis*) since 2001, most of them in Florida. See "List of Fatal Alligator Attacks in the United States," Wikipedia, https://en.wikipedia.org/wiki/List_of_fatal_alligator_attacks_in_the_United_States.

In something of a surprise, though, Erickson and his colleagues found that bite force was solely dependent on body size. "We got the same regression lines pound for pound," he told me.

In other words, if you had a crocodile, alligator, and caiman, each weighing a hundred pounds, their bite forces would be the same. All the smaller species had smaller bite forces. The seventeen-foot saltwater crocodiles generated a bite force of thirty-seven hundred pounds, but when those numbers were scaled up to the historically recorded twenty-three-footers, Erickson said, "Seventy-seven hundred pounds is not infeasible."

There were, however, two exceptions to the size/bite force correlation: the two species of gharials, whose long, skinny snouts look oddly out of place attached to a body that can reach twelve to fifteen feet in length or more and weigh in at up to two thousand pounds. Their extremely elongated jaws are equipped with 110 interlocking needlelike teeth, and the whole setup is wonderfully well adapted for slashing through the water with little resistance. But their bite force is significantly below expected values for critters of that size.

Erickson believes the gharials' specialized fishing tackle is the cause, and that it resulted in an evolutionary trade-off in which greater bite force was sacrificed for the sake of rapid fish-snatching ability, made possible by an extremely long set of toothy jaws.

Unfortunately, the two living species of gharials are critically endangered. Within the gharials' riverine habitats on the northern Indian subcontinent, their numbers may have fallen to levels measured in the hundreds of individuals.

Except for the narrow-nozzled gharials, all crocodilians, no matter their size, come equipped with some seriously powerful jaws. Erickson hypothesizes that this adaptation evolved in ancestral crocodilians during the Age of Dinosaurs, enabling them to carve out an ecological niche along the water's edge that they've successfully held for over a hundred million years. He compared crocodilian diversity to starting out with a big, powerful engine, then tweaking the attachments you could add to that megaforce-generating machine—tweaks that would include variation in the length, width, and shape of the "stuff out in front of the eyes" (Erickson's term for the jaws and teeth). These attachments helped the different crocodilians become better adapted to prey on a variety of creatures, from mollusks to fish, and from birds to big game.

When considering the bite of a crocodilian, Erickson stressed the fact that just as important as the total force a crocodilian jaw might produce is the surface area where that force is being applied—in other words, the force per unit area, or pressure. That's because this measurement not only factors in the forces generated but also the shape of the tooth. Erickson compared a pointy tooth like a crocodile canine to a shoe with a stiletto heel, which he described as more capable of damaging a wooden floor than a shoe with a flat sole. The force applied to the ground by the high-heel wearer is distributed across a smaller unit of area (the tip of the high heel) than it would be across the broad, flat sole of a sneaker.

In pointy canine teeth, the bite force is distributed over a small surface area at the tooth tip, making them effective for piercing a prey's skin or hide. Conversely, flat teeth, like premolars and molars, distribute bite forces over a greater surface area, making them ideal for turning large chunks of food into smaller chunks—a stage known as mechanical digestion. Using the tongue to smash softer food up against the hard palate (i.e., the roof of the mouth) can also be part of the process of mechanical digestion, as can the churning and mixing that takes place in the stomach.

But digestion is a two-stage process. The smaller, saliva-moistened chunks now have a greater total surface area than the larger chunks did, thus providing more space for digestive enzymes to latch on and crank up the stage known as chemical digestion (which takes place in the stomach and intestines). Eventually, the large molecules making up the small chunks get broken down into progressively smaller and smaller molecules (e.g., proteins into amino acids). The end results of chemical digestion are substances that can be utilized by cells for processes like growth, repair, and reproduction.

Of course, I was interested in the greatest bite force ever generated by an animal, extinct or extant, so I asked Erickson. He told me that was a tough question (presumably in part due to the scarcity of researchers during the Age of Dinosaurs) but that data points to a pair of prehistoric apex predators.

The first is *Deinosuchus*, a now-extinct crocodilian relative of modern alligators. *Deinosuchus* lived during the Late Cretaceous period (around 82–75 mya) in what is now the United States, where it reached lengths of almost thirty-three feet. "I don't think that any animal that has ever lived could have broken the grip of *Deinosuchus*," Erickson told me.

"So, what about escaping the grasp of an adult American alligator nowadays?" I wondered. "How hard would that be?"

"The bite force of a very large alligator is about three thousand

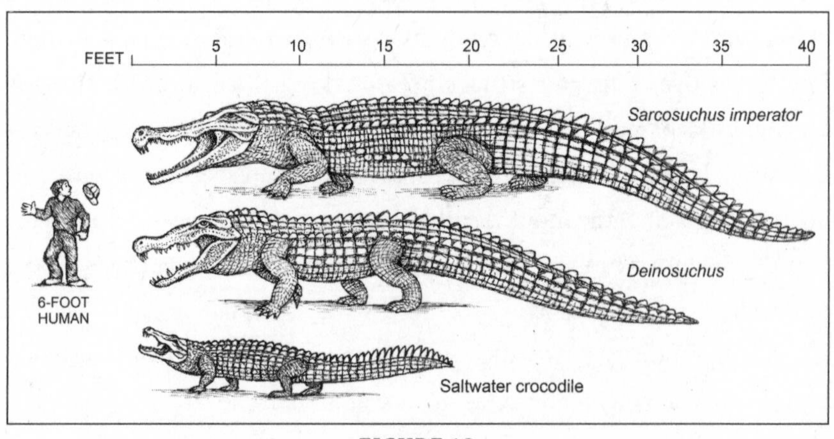

FIGURE 18

pounds, about the weight of a small car," Erickson said. "So if you can bench press a car, you are good to go . . . If not, you're lunch."

The second contestant in the Greatest Bite Force of All Time Contest is *Sarcosuchus.* With a body length of about forty feet, this behemoth lived in what is now South America and Africa during the Early Cretaceous period (133–112 mya). *Sarcosuchus* is classified as a crocodyliform (a "crocodile-like" creature). This means that although it certainly had the look of a card-carrying crocodilian, *Sarcosuchus* is not believed to be an ancestor of modern crocodiles and their relatives.* Still, Erickson believes that, like *Deinosuchus*, *Sarcosuchus* was generating a bite force of twenty-thousand pounds, a number that his team estimated by scaling up the data from extant crocodilians (see figure 18).

"I think they were right up there with the most forceful biters in history," he told me.

There are apparently limits, though, on the amount of bite force that can be generated. These relate to how much stress can be placed on the enamel covering of a tooth before it shatters. But, Erickson reminded me, this type of creature has a safeguard against that potential disaster. "Crocodilians all break their teeth," he said. "But they have one advantage over mammals—they can replace their teeth throughout their lives."

I'd been intrigued by several popular articles with titles like "The Top 10 Strongest Bites in the Animal Kingdom" and "10 Most Powerful Bites on the Planet," and I asked Erickson what he thought of those claims. Clearly treading carefully, he said that he and many other researchers

---

* *Purussaurus*, a South American relative of the caiman, which lived from the Middle to Late Miocene, may have been larger than both *Sarcosuchus* and *Deinosuchus*. Based on a single "nearly complete" fossil skull, researchers estimated body length (forty-one feet), weight (more than eighteen thousand pounds), and bite force (approximately seven tons) for *Purussaurus brasiliensis*, admitting that the "fossil crocodilian can still be considered very poorly understood." T. Aureliano et al., "Morphometry, Bite-Force, and Paleobiology of the Late Miocene Caiman *Purussaurus brasiliensis*," *PLOS One* 10, no. 2 (2015): e0117944, https://doi.org/10.1371/journal.pone.0117944.

had developed various models to estimate the bite forces in a variety of animals, but that care should be taken when interpreting the resulting numbers, primarily because of the assumptions that had gone into these models. For example, researchers using a computer model determined that a twenty-one-foot great white shark (*Carcharodon carcharias*) would generate a bite force of approximately four thousand pounds per square inch, but actual measurements were never obtained. So scientists can really only make educated guesses about the bite forces of various creatures.

"I'm pretty confident in the crocodilian [data]," Erickson told me, although he expressed skepticism about some of the other bite-force claims, such as those for *Tyrannosaurus rex*.

One confounding factor is that while ancient crocodilians appear similar to their larger modern counterparts (leading to reasonable estimates of their bite forces), there are no such extant creatures resembling *T. rex* that can be used for comparison. Still, though, Erickson speculates that the King of the Dinosaurs had a bite that was considerably less powerful than that of *Deinosuchus* and *Sarcosuchus*.

I next asked about the bite forces generated by mammals.

Erickson explained that a significant factor preventing the accurate measurements of maximum bite forces in mammals was the fact that most of them have but two sets of teeth over their lifetime. With their diphyodont dentition, mammals have evolved to be far more concerned about breaking their teeth than polyphyodont animals, which exhibit a continuous supply. Consequently, this becomes a serious issue for researchers attempting to get mammals to chomp down on something hard and potentially tooth-shattering, like a bite bar. To put it bluntly, they won't bite.

As it turns out, the key difference between the bites of mammals and crocodilians is the presence in mammals of an extremely efficient reflex arc. (See figure 19A.) You can get an idea of how reflex arcs work by asking someone you don't like to place their finger onto a hot lightbulb. The pain receptors in that person's finger quickly send a signal (via a nerve impulse traveling along sensory nerve fibers) to the spinal cord.* There, the neural signal is quickly processed (without the need for time-consuming things like conscious decision-making). Almost instantaneously, another nerve impulse (this one traveling via a motor nerve) leaves the spinal cord and heads for muscles whose subsequent contraction swiftly moves the volunteer's finger away from the heat. Only then do other sensory signals reach the brain, with the result being the perception of pain, thoughts about why that person agreed to the experiment in the first place, and the appropriate curses to vocalize their displeasure. Meanwhile, the rapid response of the reflex arc has saved that person from seriously burning their finger, all—and here's the important thing—without waiting around for conscious input from the brain.

In much the same way, when we bite down on something that's really hard, sensory receptors called mechanoreceptors associated with the periodontal ligaments initiate the so-called jaw-opening reflex. (See figure 19B.) The body's rapid response to the sensory signal travels from the nearby trigeminal nerve to the biter's jaw muscles.† As a result, the muscles that close the jaw, like the masseters, are inhibited from contracting, while the jaw-opening digastric muscles are stimulated to contract, thus preventing the broken teeth that might result from chomping down on a hard object.

---

* You'll recall that sensory nerve fibers carry nerve impulses from sensory organs (like the eyes) and sensory structures (like pain or pressure receptors) toward the central nervous system (i.e., the brain and spinal cord), while motor nerve fibers carry neural signals away from the CNS, to so-called effectors (i.e., muscles or glands). Once stimulated, the muscles or glands respond to the signals by contracting or releasing hormones, respectively. Many nerves (like the trigeminal nerve) carry both sensory and motor fibers. These are called mixed nerves.

† The trigeminal nerve innervates muscles of the face and jaw and is responsible for biting and chewing.

Because of the bone-crushing aspects of their diet, hyenas and lions are often listed among nature's most forceful biters. But after doing bite-force experiments on spotted hyenas (*Crocuta crocuta*), Erickson said researcher Wendy Binder told him that the study animals "just wouldn't bite the thing with full force . . . though one animal got really mad and

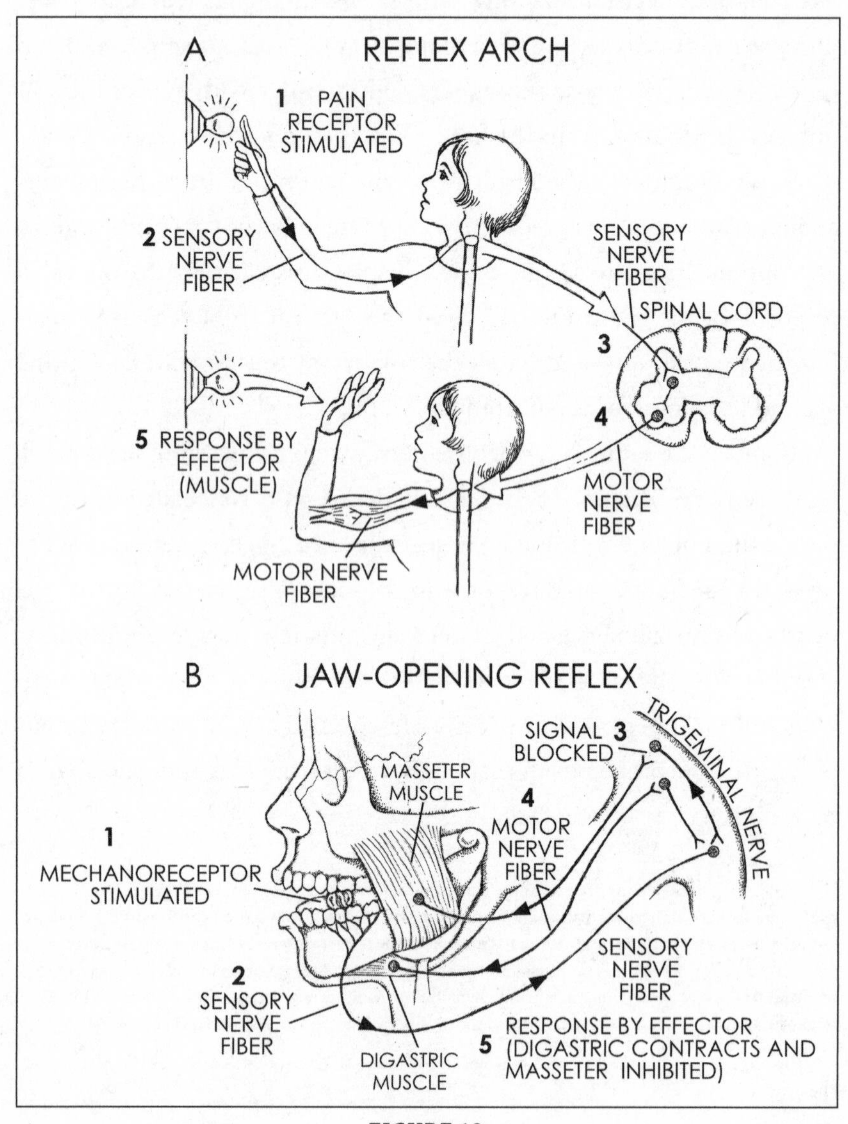

FIGURE 19

tore into it one day." It appears, however, that although their tooth-saving reflexes have prevented definitive measurement of their maximum bite forces, the bone-crushing bites of carnivores like hyenas and lions can generate pressures of between one thousand and fifteen hundred pounds per square inch.

Luckily for Erickson and his colleagues, the reflexive safeguards built into the mammalian jaw are less well-developed in the crocodilians. As a result, his toothy test subjects simply hammered the bite bar during tests, their teeth tearing through the leather and into the metal bar located below it. In the wild, of course, these teeth would have been piercing hide and muscles, clamping down upon and crushing the bones of the often-sizable prey.

"Mammals will not bite a bite-force meter with full force," Erickson said. And it's the same for birds and even lizards. "If it's a hard object and they sense it, they will not bite."

In other words, if you happen to be interested in measurements of *maximum* bite force, stick to the crocodilians!

PART II

Who, What, Where, and How Many?

[9]

By the Teeth of Their Skin

Being a fish out of water is tough, but that's how you evolve.

—KUMAIL NANJIANI

WHILE TEACHING EVOLUTIONARY biology to science majors for nearly thirty years, I learned that one key to helping my students to see the big picture was the concept of evolution as tinkerer, not creator. The basic premise is that the process of evolution doesn't come up with new stuff anywhere near as often as it tweaks the stuff that was there already. The Nobel Prize–winning French biologist François Jacob put it far better than that in a 1977 paper on the topic:

> The tinkerer gives his materials unexpected functions to produce a new object. From an old bicycle wheel, he makes a roulette; from a broken chair the cabinet of a radio. Similarly evolution makes a wing from a leg or a part of an ear from a piece of jaw. Naturally, this takes a long time. Evolution behaves like a tinkerer who, during eons upon eons, would slowly modify his work, unceasingly retouching it, cutting here, lengthening there, seizing the opportunities to adapt it progressively to its new use.

Jacob cited additional examples, one of which was the evolution of lungs. These structures and their key function, the exchange of gases between the atmosphere and the blood, apparently had their origins in the anatomy and behavior of a relatively small group of ancient lobe-finned fish known as sarcopterygians.* Approximately half a billion years ago, these critters were happily swallowing air to fill their swim bladders. These gulp-filled gas bags, connected to the gut, kept the fish from sinking due to gravity, a force that's reduced in water but still present. It's the same principle that allows inflatable "swimmies" to keep kids afloat—albeit without the young 'uns having to swallow their water wings.

Eventually, some sarcopterygians moved from salt water into swampy environments. There, they could use their fleshy lobe fins to paddle around the shallow, muddy water. The reasons for the move might have been to avoid competition for resources, since there were nutritious plants and invertebrates to eat in the swamp, and no other vertebrates around to fight over them. Alternately, the initial move might have been to escape predators, which couldn't pursue their prey into the shallows. Most likely, it was a combination of both—and likely some reasons we haven't considered yet. But whatever the reasons, these swamp-dwelling fish underwent an evolutionary tweak.

It may have been as simple as a mutation that gave them a denser concentration of blood vessels supplying their already balloon-like swim bladder. This tinkering, by random mutation, would have enabled these creatures to supplement the oxygen they were already getting from their gills—a necessity in oxygen-poor swamp water. That additional oxygen would have come from the gulped-down air in their swim bladder diffusing into the tiny blood vessels surrounding the organ. The now-oxygenated blood would then move on its way, releasing oxygen

* From the ancient Greek (*sarx* = "flesh" and *ptérux* = "wing" or "fin").

to the tissues of the body in much the same manner that the blood of air-breathing animals does today.

So, with far less creating and far more tweaking, the swim bladder had in effect become a lung—a structure originally functioning in buoyancy, modified and co-opted by evolution for a key role in an air-breathing respiratory system. Eventually, this would enable the fish's descendants to use their paddle-shaped fins—stubby, strong, and already present—for a new function: to make short trips onto land, an environment surrounded by air.

As we begin to explore the origin of teeth, you'll see that similar examples of evolutionary tweakage occurred quite frequently. Examples like these are a great way to dispel common misconceptions about evolution, which usually run along the lines of "Fish needed lungs to make the move onto land . . . so they evolved them."

First, there is no *choice* here. Adaptations like lungs and teeth can't be ordered up just because creatures happen to need them. Mutations just simply might not occur, and furthermore, not all mutations are beneficial—in fact, most are either detrimental or have no significant effect. In other words, mutation is not assured, and when mutations do occur, they do not ensure adaptation.

But before some lobe-finned fish began making abbreviated jaunts onto land, they were equipped with additional structures that had previously evolved in their even more ancient ancestors. These structures were teeth, and like lungs, they held more obscure jobs before becoming famous.

Researchers believe that the earliest fish not only lacked teeth, but also structures that evolved along with them: jaws. Consequently, the early jawless (and toothless) vertebrates, a group known as the agnathans,*

* "Agnathan" is from the Greek *a* ("without") + *gnathos* ("jaw").

were probably filter feeders—obtaining food by sifting small organisms and bits of organic matter from the water they drew in through their open mouths while swimming.

There are extant jawless fish—the eel-like lamprey and the hagfish—that serve as examples of the different ways that some vertebrates get by just fine without actual teeth. In lieu of real choppers, both lampreys *and* hagfish evolved sharpened cone-shaped projections on their tongues, known as denticles. These mounds are composed of keratin, the same hardened protein found in fingernails, hair, horns, and the sandpaper-like surface of a cat's tongue. So while hagfish don't have teeth, like most of the toothless critters that may appear to have conned their way into this book, they do have toothlike structures. And although they don't jut from jawbones or the gums covering them—a characteristic found in sharks—some of nature's "false teeth" are made of similar materials, and even look like the genuine articles.

It would be a serious mistake, though, to think of hagfish and lampreys as living examples of how ancient jawless and toothless fish were making their living. The ones alive today exhibit highly specialized lifestyles, very different from what researchers believe their ancestors were doing nearly half a billion years ago—namely, filter feeding on tiny bits of organic stuff.

Nowadays, around half of the thirty-eight species of lampreys are ectoparasites,* attaching their round, suction-cup mouths onto the bodies of other fish. (See figure 20A.) Once secured by suction and hooks of keratin, they employ a tonguelike structure covered in sharpened denticles and resembling a coarse-looking wood file known as a rasp. These are used to drill their way through scales, skin, and muscle. Survivors of

---

* Ectoparasites, like ticks, attach themselves to the outer surface of their hosts. Endoparasites, like tapeworms, live inside their hosts.

such rude attacks are often left with a circular, lamprey-mouth-sized scar to mark the occasion.

Although lamprey attacks on human swimmers are rare, long-distance swimmer Christopher Swain described just such an attack by a sea lamprey (*Petromyzon marinus*) to British television presenter Jeremy Wade. Wade later redefined the term "Don't try this at home" by prompting one of the creatures to securely attach itself to his own neck.

Since lampreys actually feed on a combination of flesh, blood, and other body fluids, they are not considered to be true members of the vertebrate vampires, a far more exclusive club whose only members are the vampire bats.

The roughly eighty species of extant hagfish feed on the bodies of dead or dying vertebrates that have settled to the ocean floor. (See figure 20B.) Since hagfish have no jaws (and so no teeth), biting off a chunk of flesh takes some doing. The process begins with the protrusion of a tonguelike blob of flesh that folds outward like a pair of grotesque flower petals. Each fold is equipped with two rows of pointy toothlike keratin. These denticles enable the hagfish to burrow headfirst into a (hopefully) deceased meal, which is often consumed from the inside out. The tonguelike flesh folds are then used to "bite" into the carcass. Lacking the jaws with which it could clamp a chunk of corpse, the hagfish ties its tail into a simple knot.

This is accompanied by the simultaneous release of copious amounts of the superslippery, fiber-thickened goo that has earned hagfish the equally pleasant moniker of "slime eels"—though they are not closely related to true eels, which have jaws. One function of this goo is to clog the gills of potential predators. Here, though, the slime assists in feeding by permitting the tail knot to quickly slide forward until it loops around the creature's head. Pushing the knot against its prey to gain leverage, the hagfish pulls the creature's head out of the loop, which causes the opposing pairs of "tongue folds" to pinch together, jawlike. The result of all these contortions is the removal of a small hunk of decaying meat, which is quickly swallowed.*

If their mealtime contortions weren't enough to cement their place in the Funky Feeding Habits Hall of Fame, hagfish can also obtain nutrition by simply hanging out inside a cadaver, where they can absorb nutrients directly through their skin. Thus, their body wall functions as a digestive organ. Hagfish are prized by South Koreans for their scaleless skins, which are fashioned into "eel skin" belts.

Some early anglers used the hagfish's unusual feeding habit to perfect a method of capturing them that would never be confused with fly-fishing. All they required was a long rope, some weights, and a marker buoy—oh, and a dead cow (or something similarly large and equally dead). I invite readers to use their imagination as to how after a couple of days on the seafloor, this particular combination of bait and tackle would have rewarded anglers with a deck full of bloated bovine, uniquely animated from within.†

The feeding habits of lampreys and hagfish (ectoparasitism and necrophagy, respectively) allow them to avoid competing with jawed fish—who have clearly evolved much cooler ways to obtain their daily nutritional requirements. And though a relatively few jawless species survive today, during the Devonian period (419–359 mya) widespread competition and other factors related to changing environments drove

---

* In July 2017, a truck carrying a large cargo of live hagfish (don't ask) braked hard to avoid slamming into a construction zone. Unfortunately, the driver had apparently overlooked one of the cardinal rules of slime eel transport: securing them safely. As a result, thirteen containers, weighing a total of seventy-five hundred pounds, spilled onto Oregon's US Route 101. One of these containers struck a car, causing a four-vehicle pileup, which, thankfully, resulted in only minor injuries. Finding themselves suddenly strewn across a major roadway, the stressed-out hagfish did what hagfish do; they produced slime—hundreds of pounds of it. A video posted by a local fire department shows the creatures "croozing" on down the highway, in a bid to escape the accident scene and return to the Pacific Ocean. Tragically, there were no hagfish survivors, and the deceased slime eels and their namesake secretions were removed with the aid of fire hoses and construction equipment.

† In more recent times, olive or pickle barrels filled with fish scraps and equipped with a one-way entrance were substituted for rope-tethered Elsies.

most of these fish to extinction. Their demise shows clearly how evolutionary innovations (here, adaptations like jaws and teeth) can lead to an explosion in diversity (i.e., more and more species) within the groups that possess them, while it's curtains for those that don't.

So, HOW DID jaws and *real* teeth evolve? While there is general agreement about how jaws came about, there are two current hypotheses regarding the origin of teeth, though both may turn out to be correct.

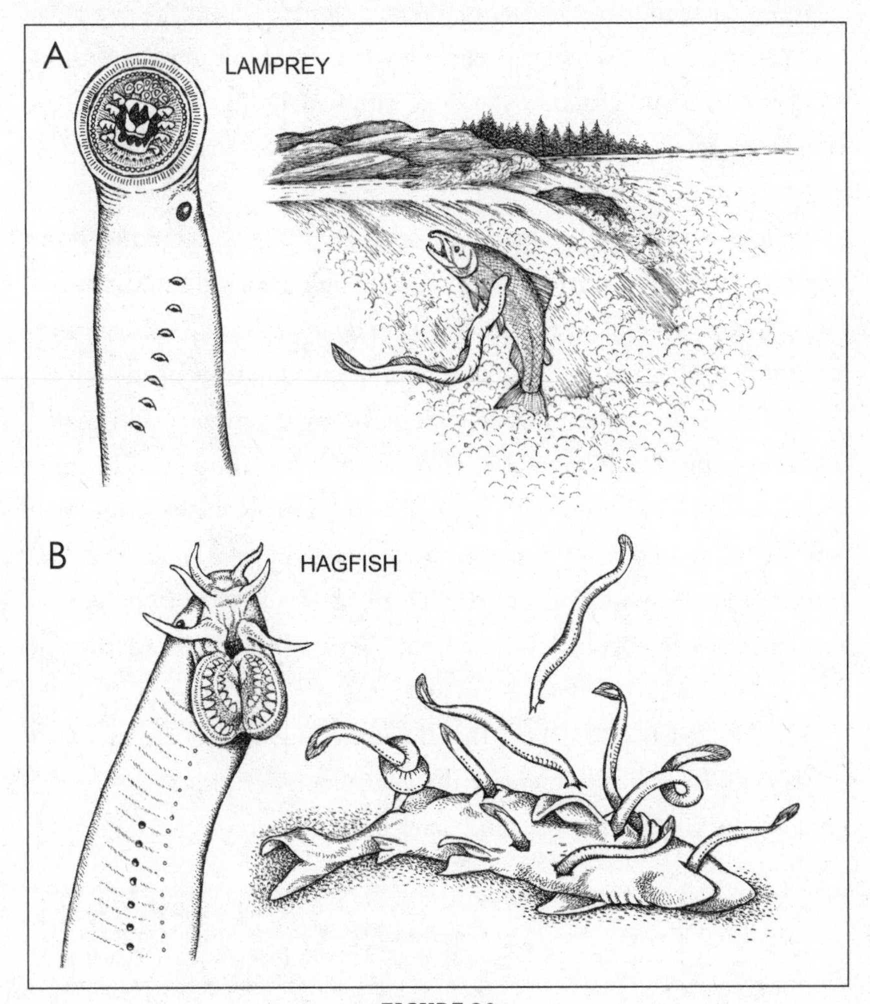

FIGURE 20

First off, and very much in line with the concept of evolution as tinkerer, researchers believe that jaws evolved in ancient fish, from previously existing frameworks of cartilage or bone, known as gill or branchial arches. These develop in all vertebrate embryos, though their number and fate vary. In fish, gill arches are paired and roughly boomerang-shaped. They function by supporting the feathery gills involved in respiration. Both the gills and gill arches are situated in a space called the opercular (or gill) cavity. This is located in back of the oral (or buccal) cavity—which, in turn, sits behind the mouth. It is within the opercular cavity that the gas exchange occurs between the incoming oxygen-rich water and the tiny capillaries located within the gills. The oxygenated blood then leaves the gills via additional blood vessels before heading off to supply the body.

By now, you might be wondering what all of this has to do with the origin of jaws and teeth. Not to worry, though, because the structures formerly known as gill arches (two pairs of them anyway) took on brand-new roles. And once they did, the vertebrates would never be the same.

Although there is some debate over the original number of gill arches, researchers think there may have been as many as nine pairs in some jawless ancestral filter feeder. (See figure 21A.) After the anterior-most pairs of arches were lost (for reasons that remain puzzling), the next two pairs in line (the mandibular and hyoid arches) underwent what is arguably the most important evolutionary makeover in vertebrate history. They became jaws. (See figures 21B and 21C.)

Paleontologists also believe that the initial function of jaws was not to grasp and bite but to increase the efficiency of respiration by opening and closing the mouth.* Only later did jaws evolve into structures

---

* Though the system was far simpler in ancient fish, eventually changes in the volume and pressure of the oral and opercular cavities would allow large volumes of water to be pumped in through the mouth and out through the gill slits located adjacent to the gills. Basically, more efficient processing of incoming water meant more oxygen for the fish, which enabled them to evolve larger, faster, and more complex bodies.

that were also employed during feeding. Eventually, they would coevolve with teeth—which also appear to have served different roles before landing their more famous gig. Evolution as tinkerer, not creator, strikes again.

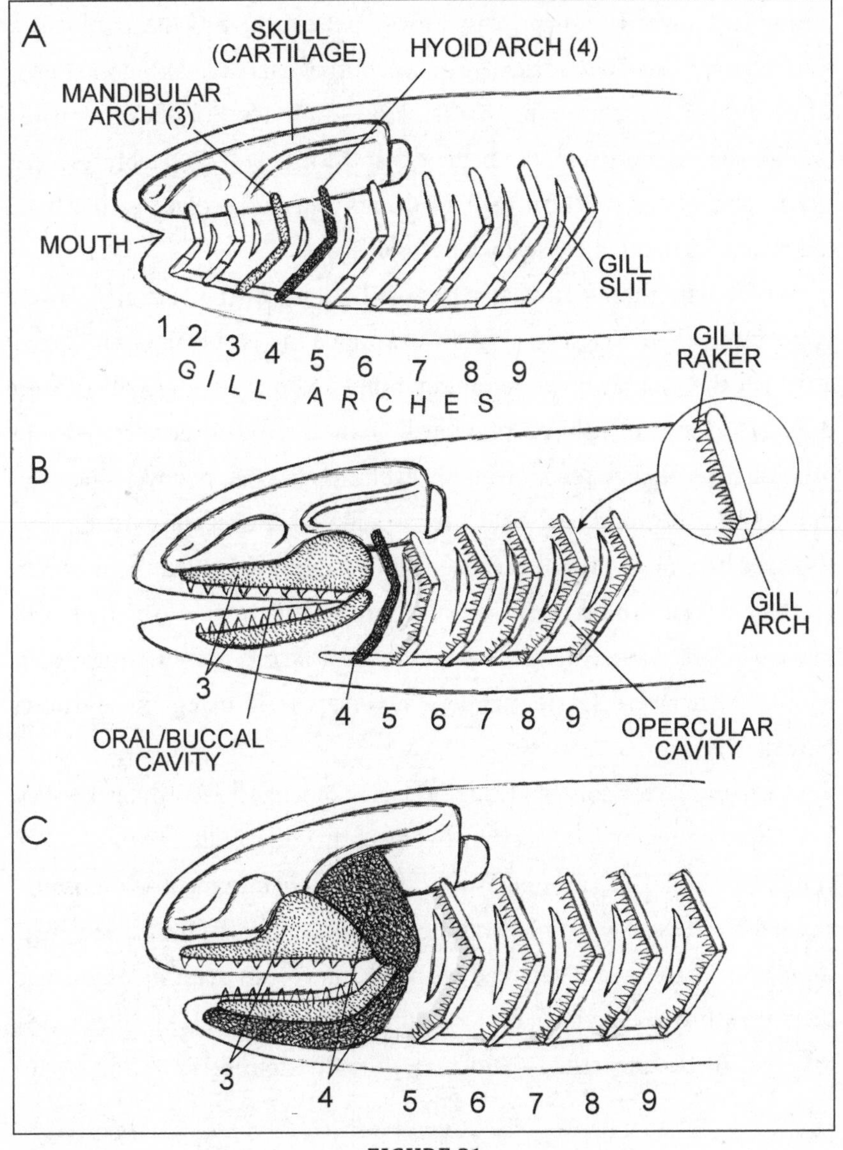

FIGURE 21

THE NEXT QUESTIONS: Who were the first jawed fish, and when did the first teeth show up?

I asked Per Ahlberg, paleontologist and professor at Uppsala University, who told me that the earliest probable jawed vertebrate remains are scales dating to the beginning of the Late Ordovician period (around 458 mya). In a surprise to the researchers, though, the fossil scales likely came from a chondrichthyan, a group of cartilage-skeletoned fish, which currently includes the sharks, skates, and rays. But like other paleontologists, Ahlberg does not believe that the first jawed vertebrates were sharks or shark relatives. Instead, he thinks that jaws evolved earlier than the creatures that left those ancient fossil scales.

As for why there's currently no fossil record of the very first jawed vertebrates, there are several likely reasons. One is the apparent speed at which the first jaw bearers rapidly branched off (i.e., diversified) into different groups of fish. As previously mentioned, this accelerated rate of speciation makes sense from an evolutionary perspective, since the first jawed fish would have held a remarkable edge over the existing jawless species. Mainly, that's because jaws would have helped the new fish on the block to exploit their watery environments in new, different, and more efficient ways. This is comparable to a farmer moving into town with a John Deere tractor while everyone else is using horse-drawn plows.

A second factor contributing to the lack of a fossil record for the very first jawed vertebrates is that they lived nearly *half a billion years ago.* For comparison, that was around 250 million years before the first dinosaurs arrived on the scene. Combining the speed at which the first jawed creatures appear to have diversified with the enormous span of time since these creatures existed, it's understandable that their fossilized remains have yet to be unearthed—and may never be found. Rare, but better represented in the fossil record are the species that evolved not long after the first jaws evolved.*

The prime players in the story of tooth origins are an extremely successful group of armor-plated fish known collectively as placoderms.† A caveat, though, is that there has been some late-developing controversy about where teeth came from. It involves a mixed bag of ancient jawless fish—some with bodies that were never preserved in the fossil record.

First up, though, the placoderms. In something of a familiar refrain, once these vertebrates appeared on the scene, their jaws and paired fins gave them a huge competitive edge over the jawless fish that preceded them. Equipped with these adaptations and more, the placoderms diversified into an array of creatures exhibiting different sizes, shapes, and habits, and they would come to dominate the sixty million years spanning the Age of Fishes.‡ Known today from approximately 350 genera preserved in the fossil record, placoderms were divided into two groups: the arthrodirans, which included the thirty-foot Devonian sea monster *Dunkleosteus* (see figure 22A), and the acanthothoracids, which were far smaller and less diverse, but ultimately no less important to the paleontologists who studied them (see figure 22B).

The good news is that the placoderms were a very early branch of the jawed vertebrate tree and some of them had true teeth. The bad news is that nobody can pinpoint exactly which species was the first to have

---

* Something very similar may have happened after the evolution of bats, the first flying mammals. Like the appearance of teeth nearly half a billion years earlier, the ability to fly likely led to an explosion of bat diversity, making the search for the first bat a difficult, and perhaps impossible, task.

† Greek for "plate skin."

‡ Those of you who may be feeling badly about the placoderms driving most jawless fish to extinction might take heart at the fact that the two existing groups of fish: the chondrichthyans (sharks and their relatives) and the osteichthyans (bony fish) set the bulky-armored placoderms on their own path to Fossilville by the end of the Devonian period.

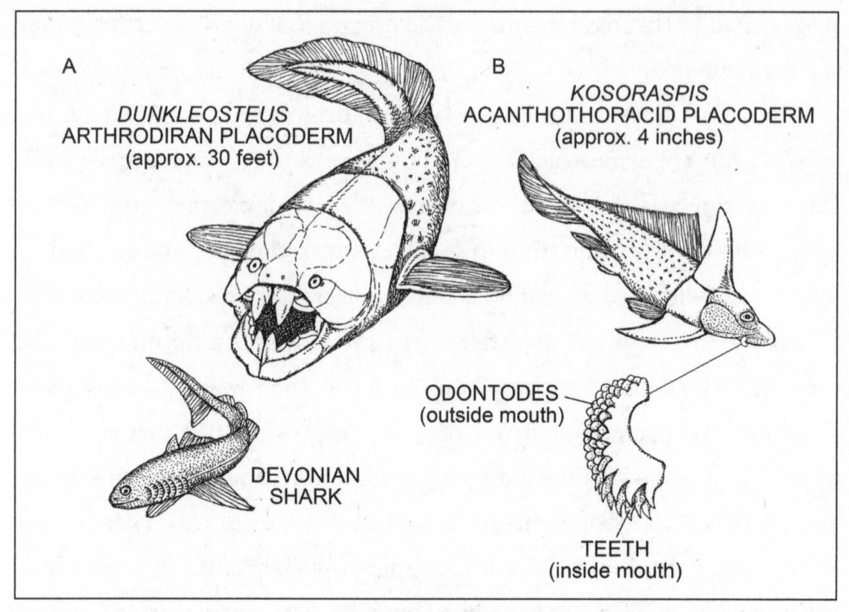

FIGURE 22

teeth—though unlike the first jawed vertebrate, researchers now have a pretty good clue.

Interestingly, the giant predator *Dunkleosteus*, whose massive skull currently threatens visitors of all ages as they enter the American Museum of Natural History's Hall of Vertebrate Origins, did not have teeth. Instead, this late arrival to the Age of Fishes wielded upper and lower jawbones with wickedly sharpened edges. When the jaws closed, these self-sharpening blades slid past each other, slicing off chunks of whatever unfortunate prey happened to be in the vicinity—including smaller *Dunkleosteus* individuals!

Until recently, there was little controversy over where the first teeth came from. Beginning with a series of classic nineteenth-century papers written by some of the first paleontological heavyweights—like comparative anatomist/paleontologist Richard Owen (he of *Hyracotherium* fame) and biologist/geologist Louis Agassiz (1807–1873). Agassiz, a devout creationist, had described similarities between fish dermal denticles and vertebrate teeth. Owen recognized the same similarities. By the 1960s and through the '80s, researchers were voicing the connection loud and clear: the first teeth had likely evolved from previously existing structures on the outside of some ancient fish bodies.

Eventually, the list of tooth precursors would expand to include the bony bumps called odontodes found on the body surfaces of some placoderm fossils. For those keeping track of such things, odontodes differ from the bumps also known as denticles, seen in hagfish and lampreys, which are made of keratin—the fingernail stuff, not bone. Unfortunately, the terms are sometimes used interchangeably, which can confuse the issue.

An important concept related to tooth origins is that during development, bones can form one of two ways. Endochondral bones develop from cartilage into long bones like the femur, the tibia, and the humerus. Dermal bones form within the skin (not from previously existing cartilage). These flat bones can be found in places like the walls of the skull, the jaws, and the scapula (shoulder blade). Dermal bone also makes up fish scales and odontodes—which is precisely where most researchers now believe vertebrate teeth originated.

Strengthening the hypothesis that teeth evolved from structures on the exterior of the body was the discovery that some of them were remarkably toothy in composition. In what became known as the "outside-in hypothesis," it appears that over time, odontodes that were originally located on the surface of the skull of some ancient placoderms gradually migrated into the mouth and onto the jaws. Once relocated, they took on new roles as teeth. The outside-in hypothesis became the go-to explanation for the origin of teeth for half a century. What was missing, though, was fossil evidence of the proposed movement from odontodes, outside, to teeth, inside.

Then, in the 1990s, researchers like Moya Smith and Michael Coates came up with an alternative hypothesis. They proposed that the discovery of toothlike structures in two groups of ancient jawless vertebrates, the conodonts and thelodonts, pointed to an earlier tooth origin than that suggested by Team Outside In. Smith and her colleagues reminded everyone that both of these groups existed *before* the placoderms, which had yet to demonstrate (within the fossil record, at least) the proposed outside-in transition from odontodes to teeth. They also asserted that the formation of teeth required the presence of endoderm, an embryonic tissue which was not found in the skin—and which was exactly where Team Outside In claimed the first teeth had originated.

Countering century-old dogma, the "inside-out hypothesis" was born. Its supporters proposed that teeth had evolved from odontodes located in the pharynx, the short, vestibule-like space in the rear of the mouth (commonly referred to as the throat). It was from the pharynx that the inside-outers hypothesized that these bony bumps migrated forward, perhaps multiple times and in different species. Eventually, and as jaws evolved (remember, they initially functioned in respiration), the odontodes would take on new roles as teeth. These researchers strengthened their argument by adding that, unlike the skin, the pharynx was a region where the embryonic endoderm they claimed to be a requirement for tooth development was found in abundance.

One inside-out scenario involves gill rakers. These are toothlike projections of varying shape and size jutting off the recently discussed gill arches. Composed of bone or cartilage, gill rakers function by preventing swallowed food particles from contacting the delicate gills anchored to the gill arches. One hypothesis is that in some ancient jawless vertebrates, as the front-most gill arches migrated forward into the oral cavity, taking on a new role as jaws, the gill rakers they bore (or similar bony bumps) evolved into teeth. (See figures 21B and 21C.)

At this point, the debate really took off, with researchers jumping in on either side of this new and unexpected controversy. What interests me most is how some seriously cutting-edge technology had been used to further explore the question of vertebrate tooth origins, in ways that nineteenth-century naturalists like Agassiz and Owen could never have imagined. I find this fascinating, and I've often advised students searching for original research projects to look for questions that were *thought* to have been answered long ago. Then, by using modern technology and tweaking those questions for the twenty-first century (rather than using the approaches that might have been taken "in the old days"), new projects can be developed. I know this technique works because it is precisely how I came up with a PhD dissertation project while working as a grad student at Cornell back in the 1990s.*

And where did this new hypothesis leave those searching for the origins of vertebrate teeth? Or, to put that more accurately, which hypothesis is currently supported by the most evidence? The answer remains tricky.

The toothlike structures seen in the ancient jawless vertebrate mouths and cited as precursors of vertebrate teeth (the inside-out hypothesis) instead appear to be examples of convergent evolution. In other words, toothlike structures evolved independently in several groups of jawless vertebrates, but they may not have evolved into vertebrate teeth. The claim that teeth cannot form in the absence of the embryonic tissue endoderm (an argument for the inside-out theory) has also been

* In my case, the question was: How can bats hang for such prolonged periods of time? The answer, a Velcro-like locking mechanism located on the bones of their toes, first described in 1903 (in a German journal and without photography). Back then, anatomy was mostly a descriptive science and anatomists were not asking late-twentieth-century-style anatomy questions. So I did—by linking these neat "digital locks" to topics like ecology, evolution, and biomechanics. By doing so, and by employing new technology (like electron microscopy), I was able to craft several papers, as well as a large chunk of my thesis on a cool topic that few had ventured to investigate further than the original researcher, Josef Shafer.

disproven and thanks to modern technology, there is clear evidence in at least one placoderm that odontodes located outside the mouth *did* grade into teeth once they were inside the mouth.

In this case, the discovery occurred in response to the challenges made by Team Inside Out. Researchers set out to examine a set of extremely delicate specimens that had been discovered a century earlier in what is now the Czech Republic. Using a high-tech form of nondestructive dissection, they showed that in one of the specimens they examined, an acanthothoracid known as *Kosoraspis*, the odontodes located around the outer margin of the mouth graded into sharpened, recurved teeth (like dog and cat canines) once they were inside the mouth. (See figure 22B.) This strengthened the claim that they were used for feeding. Just as importantly, similarities between the teeth of acanthothoracids like *Kosoraspis* and those of the bony and cartilaginous fish that followed them suggest a close evolutionary relationship—quite possibly a relationship in which the acanthothoracids were the ancestors of both of those modern fish groups.

The debate, however, is not over. In 2010, a team led by researcher Gareth Fraser proposed an "inside and out" model, in which toothlike structures could have formed wherever odontodes were present—in some groups, from odontodes on the skin; in others, from bony bumps in the pharynx. Accordingly, teeth in the vertebrates would have originated multiple times.

Before moving on, it should be noted that both acanthothoracid and arthrodiran placoderms eventually had teeth that were firmly attached to their jawbones, the condition still shared by bony fish and the terrestrial vertebrates that evolved from them. Sharks' teeth, though, are not directly attached to the jaw but are instead embedded in their fleshy gums. This tosses yet another wrench into the idea that sharks were the ancestral toothed vertebrates. Instead, it appears likely that ancient sharks were a highly specialized offshoot of the earliest jawed fish.

Unfortunately, where exactly the first chondrichthyans came from is not currently known from the fossil record. Primarily, that's because their cartilage skeletons break down more rapidly than bony skeletons. Thus, a scanty fossil record (except for teeth) has produced serious muddying at the base of the evolutionary tree from which the ancient sharks mysteriously sprang.

We do know that shark teeth are essentially identical in form to the placoid scales that cover their bodies. (See figure 23.) These tiny hook-shaped structures give sharkskin its rough, sandpaper-like texture and function as a mechanical barrier, protecting the shark from abrasion, predators, ectoparasites, and the environment.* Placoid scales are also arranged with their sharpened points facing the tail. This produces a streamlining effect, by reducing friction on the shark's body as it glides through the water.

It is thought that, like the inward-migrating odontodes that led to the first placoderm teeth, a similar migration of placoid scales in some ancient shark (or sharks) resulted in the formation of sharks' now-famous dentition.

Shark teeth, like those of other vertebrates, are also composed primarily of dentin covered by a thin layer of ultrahard enamel (or enameloid) and have an internally located pulp cavity. For the most part, though, that's where the anatomical similarities with other vertebrate teeth end and some real differences begin.

---

* In the seventeenth and eighteenth centuries, sharkskin began taking the place of untanned horsehide for use as a product known as shagreen. Often ground down to leave a slightly rough surface, its tough, waterproof nature made it an efficient cover for luggage, books, toiletry cases, etc. In its un-ground-down state, sharkskin often functioned as sandpaper.

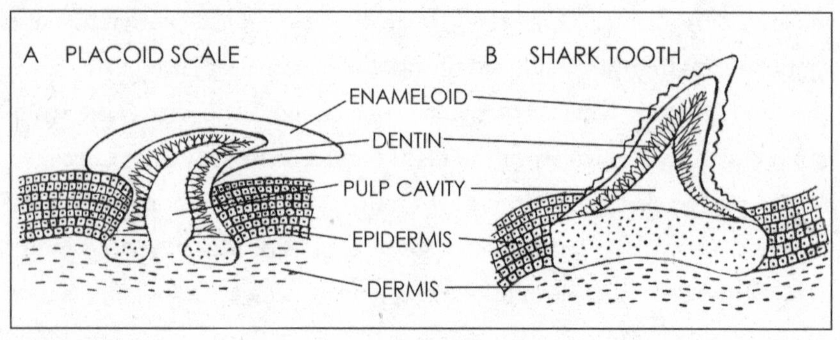

FIGURE 23

As mentioned, shark teeth are loosely embedded in the gums, and, considering the violence associated with their function, they get knocked out quite frequently. That's partly because they do not have roots, the portion of teeth that extends deep into the jawbones. They are, however, quickly replaced by teeth waiting patiently in rows behind those currently in use. With estimates of sharks losing and replacing upward of thirty thousand teeth over a lifetime, this is an extreme example of a condition known as polyphyodonty—continual tooth replacement.

Polyphyodonty is seen in only three mammals—elephants, kangaroos, and manatees—and even then, on a limited scale, since, unlike sharks for example, tooth replacement is not unlimited. But this form of tooth replacement is found in thousands of nonmammalian vertebrates, including bony fish, amphibians, and reptiles (like alligators). Here, rather than the unique condition seen in sharks, where rows and rows of teeth sit in an on-deck position, ready to spring into place, most vertebrates with the ability to regrow more than two sets of teeth do so because of the continued presence of tooth-producing stem cells in their jaws. Once an existing tooth has been worn down or is no longer present, these latent (or inactive) stem cells become activated, initiating the growth of a tooth bud (a bulb-shaped cluster of cells), which develops into a new tooth.

Now, though, our toothy story begins to transition from aquatic to terrestrial environments. During the Devonian period, some of those air-gulping lobe-finned fish we met earlier began making brief forays onto land, a journey made easier by the presence of primitive lungs. Evolving in tandem with these respiratory structures were the sturdy paddle-like fins that gave the group its name. Unlike the delicate-looking pectoral and pelvic fins seen in most bony fish, these fins were bulky, since the muscles that controlled them were located outside the body. Like other anatomical systems, however, structures associated with lobe fins underwent their own set of evolutionary tweaks. Some were lost. Others underwent modification gradually as they took on new roles as weight-supporting limbs, holding the body up to resist the force of gravity while increasing the efficiency of terrestrial locomotion. Their fishy-looking teeth underwent some tinkering as well, evolving through the process of natural selection to better fit their owners' new semi-aquatic habitats. Eventually, and over millions of years, some of these fish species would have evolved enough landlubber traits (e.g., thick skin to prevent water loss, and the loss of gills in the adult stages) that nearly half a billion years later, their brainy, two-legged descendants would christen them "amphibians."

[10]

A Painless Guide to Tooth Basics

When it lives in a reef and has two sets of teeth, that's a moray.
When the jaws open wide and there's more jaws inside, that's a moray.

—Unknown

Like a trip to the dentist, we have some potentially unpleasant stuff to get out of the way. This brief excursion should make what follows somewhat easier to swallow.

As with the basic composition of vertebrate teeth, there are similarities and differences in how teeth evolved their attachment to the jaw. Setting aside the sharks and their pals, within the remainder of the vertebrates there are three basic types of dentition, their differences relating to the way that teeth are attached to the upper and lower jawbones.

The term "acrodont" (see figure 24A) comes from the Greek word *akros* ("highest") and is here related to the fact that these nonrooted teeth are fused directly to crests on the surface of the maxillary (upper) and mandibular (lower) jawbones. Thought to be the ancestral (or primitive) condition in vertebrates, acrodont teeth are found in bony fish, amphibians, and a relatively few reptiles, like Old World chameleons. Because of the tooth's superficial attachment to the jaw (there are no roots), they can be lost during activities like prey capture or feeding. Generally, acrodont teeth are not replaced. Instead, in the older adults of some species, the crests of the jawbones themselves become sharpened and serrated. This

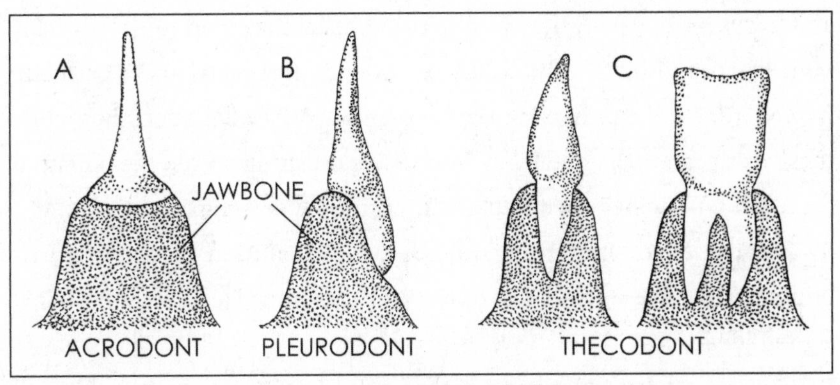

FIGURE 24

resembles the condition seen in turtles—which do not have teeth. The exposed surface of turtle jawbones is covered by sharpened ridges of keratin. This arrangement forms a birdlike beak, which functions as an efficient alternative to teeth.

Most lizards and snakes exhibit pleurodont dentition (see figure 24B), in which the teeth have roots but there are no alveoli (the deep sockets in the jawbones). The name is aptly derived from the Greek *pleura* ("side"), since the labial side (i.e., lip/cheek side) of a pleurodont tooth is attached to the lingual side (i.e., tongue side) of the jawbone. The tooth is held in place by tough fibers of connective tissue known as periodontal ligaments.* Pleurodont teeth can be shed and replaced throughout the life of their owner.

Dinosaurs, crocodiles, and mammals, including humans, have (or had) the third type of tooth attachment to the jaw: thecodont dentition (see figure 24C). Here, tooth roots are strongly fused to the upper

* If you've ever wondered about the difference between ligaments and tendons (or even if you haven't), while both are composed of tough connective tissue fibers, ligaments connect bone to bone (as in the ligaments that connect pleurodont teeth to their jawbones) and tendons connect muscle to bone (like the tendons connecting the masseter and temporalis muscles to the mandible). Since the temporomandibular joint sits between the bulk of these two muscles and their attachment sites on the skull and mandible, when the muscles contract, mechanical energy is applied to the mandible via the tendons, and the jaws close.

and lower jawbones in the alveoli. The roots are secured to the alveolar bone (surrounding the alveoli) by periodontal ligaments and a perfectly named mixture of minerals called cementum.* (See figure 25.) But while these teeth can withstand some impressive crushing power, the potential for tooth replacement in adults is either limited or nonexistent.

Because of the importance of thecodont dentition throughout this book, a brief review of its most important anatomical features is probably a good idea.

Covering the alveolar bone of the jaw and continuous with a layer of mucous membrane lining the inside of the mouth (the inner surface of your cheek, for example) is a fleshy region known as the gingiva, or gums. The gingiva also contacts the tooth near where the root begins, at a region called the tooth neck. Just past the neck, and above the gumline, is the crown, the visible portion of the tooth, characterized by a thin outer layer of enamel.

The term "occlusal surface" is used to describe the places on the crown that meet when teeth of the opposing jaws come together (or occlude). The bumps or points found on these surfaces are called cusps (from *cuspis*, Latin for "point"). These fit into cusp-shaped depressions located on the teeth of the opposing jaw, often in a mortar/pestle arrangement that allows food to be sliced up or crushed. Cusp presence, absence, number, and shape are characteristic of tooth type within an individual (canines vs. molars, for example) and also across the animal kingdom (molars in pigs vs. molars in horses). Cusps are integral, therefore, in providing researchers with information on diet, evolutionary relationships, and much more.

Deep within the alveolar bone of the jaw, tiny blood vessels pass to and from the tooth, through a hole in each root tip called the apical

---

* This type of peg-fits-into-socket arrangement is actually a specialized form of immovable joint called a gomphosis.

foramen. The arteries (with their incoming blood) and veins (containing blood exiting the tooth) pass through the root, converging in the hollowed-out portion of the tooth, the pulp cavity. It is here that materials are exchanged between the vessels and the tooth. These include gases like oxygen and carbon dioxide, nutrients, waste products, and building

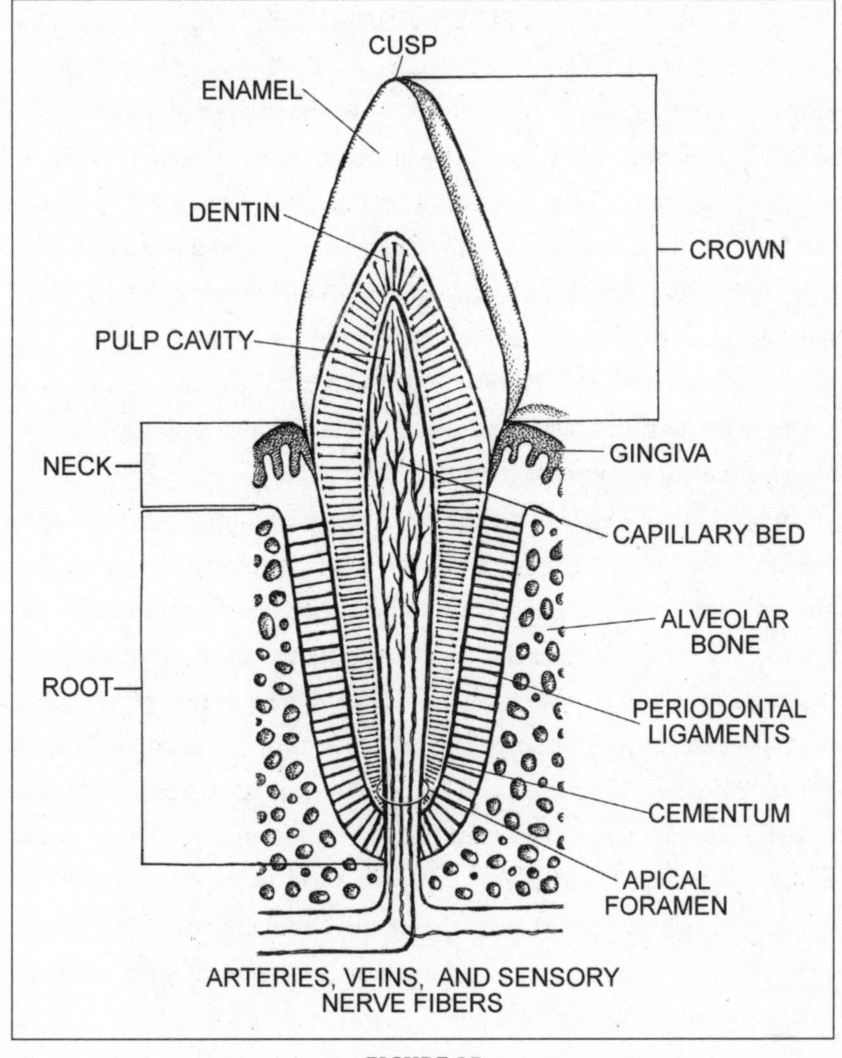

FIGURE 25

materials like minerals. Also entering the tooth through the apical foramen are tiny sensory nerve fibers, which make their presence known when dental problems like tooth decay arise. Together, the nerve tissue, blood vessels, and the jellylike tissue that surrounds them are known as the tooth pulp.

TOOTH SHAPE AND NUMBER

Homodonty, in which all the teeth have the same general shape, is also characterized by large numbers of teeth. In land mammals, the record holder for most teeth is the giant armadillo (*Priodontes maximus*), with 74 teeth, while the spinner dolphin (*Stenella longirostris*) is the toothiest of all mammals, with up to 260 sharp, pointed teeth filling its long, skinny jaw. For sheer tooth number, though, bony fish blow away the mammalian competition. Not confined to the jaws, as in mammals, fish teeth can also be found on tongues and gill arches and lining oral cavities. The toothiest fish species appears to be the lingcod (*Ophiodon elongatus*), with around 550 teeth.

In addition to a pair of normal jaws, these and other bony fish species have a set of pharyngeal jaws. Formed from gill arches, they bear pharyngeal teeth of varying shape and number. Most pharyngeal teeth function by crushing and grinding up food items before they pass on to the esophagus and stomach. Goldfish (*Carassius auratus*) possess only pharyngeal teeth (see figure 26A), while species like lingcod and moray eels have teeth on both their oral and pharyngeal jaws. (See figure 26B). In the latter, after the oral jaws secure the prey, the pharyngeal jaws rush forward to clamp down on the startled but still-living victim, which is quickly dragged backward from the mouth and delivered to the esophagus. This is thought to be an adaptation related to the limitations of feeding with a snakelike body.

Unlike the homodont dentition found in most bony fish, tooth shape can differ significantly in venomous snakes and crocodilians (though

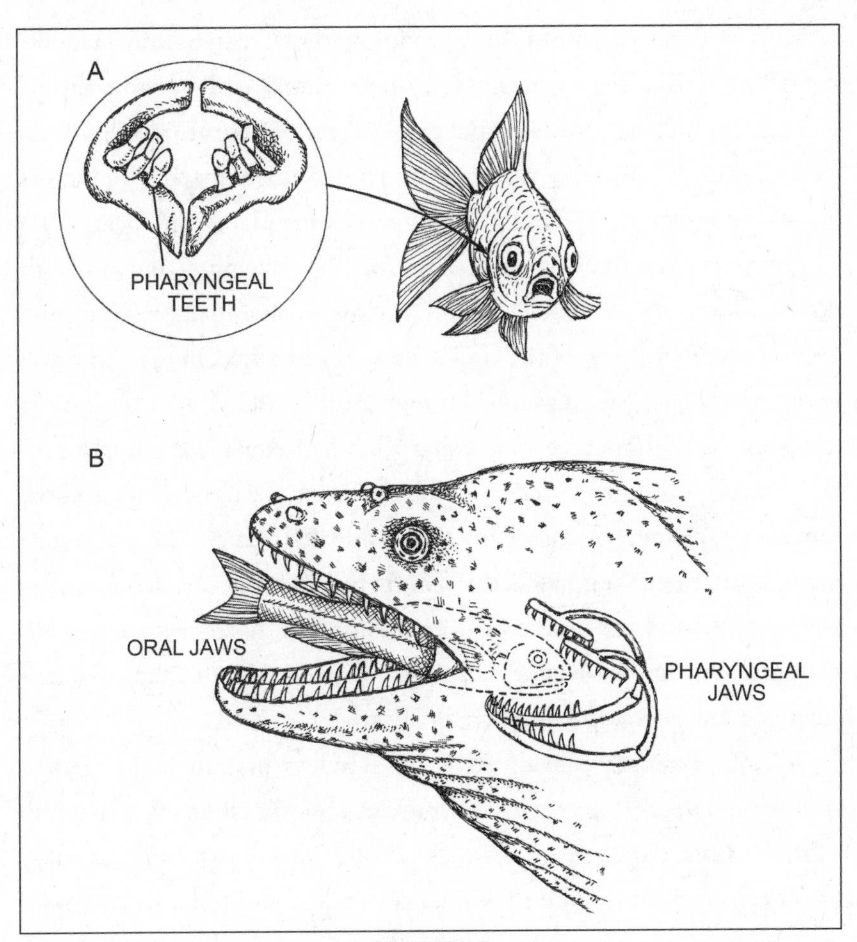

FIGURE 26

not in most amphibians and reptiles). This heterodonty is most pronounced in the mammals, though, where up to four different types of teeth often reflect a significant division of dental labor. From the anterior of the skull to the posterior (i.e., from front to back), the teeth found in typical mammalian mouths are incisors (I), canines (C), premolars (P), and molars (M).

You can think of a mouth holding this assortment of teeth as a toolbox. The tools within will vary depending on the type of food being consumed. Meat eaters, for example, require a very different set of tools than herbivores. But even within these groups, variation can be considerable. Carnivore teeth differ from those of insect eaters, and the same can be said for leaf-nibbling browsers and grass grazers. There are also omnivorous mammals, like bears, raccoons, and humans, equipped with tools (I mean, teeth) capable of obtaining and processing plants and meat.

Omnivory is actually widespread across the vertebrates, but although the tag does not necessarily denote a close evolutionary relationship, omnivores often share toothy similarities. For example, in many omnivorous mammals (e.g., pigs and humans), the chisel-shaped incisors at the front of the mouth evolved sharp beveled edges, for snipping off bite-sized chunks. Just to the rear of the incisors, the pointy canine teeth function in piercing and gripping, and while their fang-like prominence in dogs gave them their name, this feature isn't as pronounced in humans anymore. Behind the canines, the premolars and molars evolved broad, flattened surfaces, perfect for crushing and grinding food items that had been previously gripped, snipped, or sheared.

But omnivore-style teeth aren't confined to mammals, as anyone who has ever seen the remarkably human-looking choppers flashed by a group of fish commonly known as the sheepshead (*Archosargus* spp.). Upon closer examination, resemblances end—thankfully, since unlike most humans you'll run into, the sheepshead has three rows of teeth on each side of its upper jaw and two on its lower jaw.*

Besides the often-stark differences in their shape, heterodont teeth are also highly variable in presence/absence, size, and number. To help sort out all this variation and to avoid confusion, a dental formula is used to specify the number and type of teeth on one side of the upper and lower jaws. For humans this can be written I2/2, C1/1, P2/2, M3/3 or shortened to 2/2, 1/1, 2/2, 3/3. Here the formulae indicate two upper and two lower incisors, one upper and one lower canine, etc. Since dental

* So, yes, even some bony fish have heterodont dentition, though it is relatively rare.

formulae specify only one side of the mouth, to get the total tooth number you would add the numbers up and then multiply by two. In this case, that gives us a total of thirty-two adult teeth (see figure 27).

While we're on the topic, here's the dental formula of a rabbit-sized marsupial called a potoroo: 3/1, 1/0, 1/1, 4/4. See if you can figure out

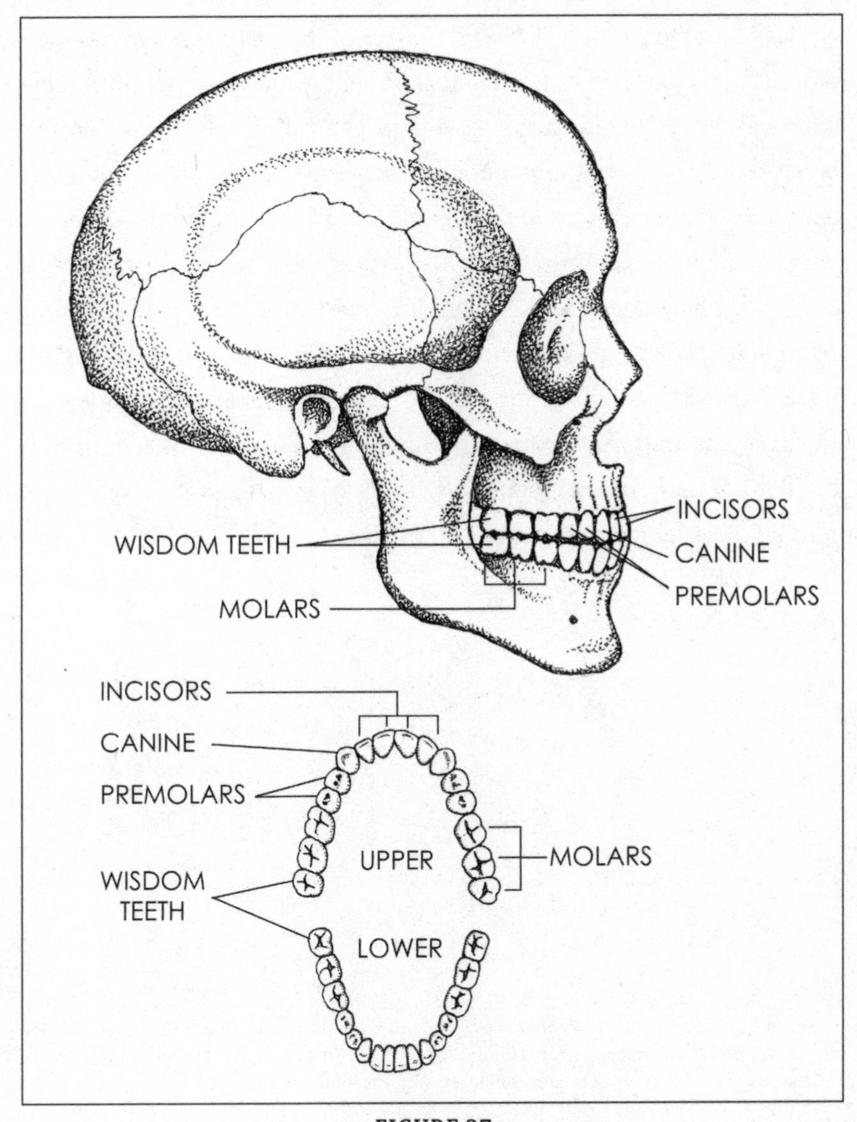

FIGURE 27

what's going on here dental-formula-wise before we move on. I'll wait.

Okay, truth be told, I also used this specific critter to demonstrate that, unlike the typical condition in humans, the number of teeth on the upper and lower jaw can vary.*

There are also many examples among the roughly fifty-four hundred species of mammals in which one or more of the tooth types mentioned above is completely absent. Rodents and rabbits lack canines, and their skulls can be readily identified by a large gap called a diastema, located between their incisors and premolars. (See figure 8, page 58.) Adult manatees have neither canines *nor* incisors, and premolars are only seen in young manatees. Reminiscent of the way elephants use their prehensile trunks, manatees use their similarly maneuverable lips to grip and manipulate the sea grasses and other aquatic plants upon which they graze. This pretty much eliminates the need for incisors and canines.

Given the importance of teeth across a wide swath of the animal kingdom, it may come as something of a surprise that a relatively large number of vertebrates are, in fact, toothless.

---

* For those of you who took "the tooth test," the potoroo (*Potorous* sp.) has three upper incisors but only one lower, no lower canines (as indicated by the 0), one upper and one lower premolar, and four upper and four lower molars. Bonus answer: the total tooth number is 30.

PART III

Teeth, Who Needs 'Em?

[11]

Of Fish and Frogs

Theories pass. The frog remains.

—Jean Rostand, *Inquiétudes d'un biologiste*

Much of what you've read so far supports the hypothesis that vertebrate teeth were a key innovation that enabled the jawed vertebrates to undergo an explosion in diversity, permitting them to feed on every conceivable form of nutrient- and energy-containing matter—from gnats to nuts. As a result, Team Backbone was able to move into, successfully adapt to, and exploit nearly every environment on the planet.

But what about the vertebrates that lack teeth, those of the so-called "edentate" variety? For extinct groups, like some dinosaurs, how did they survive and thrive for millions of years? And for toothless extant vertebrates, what is it that enables them to persist in a world of toothy competitors?

Before we explore these questions, there's a fact that makes things even more interesting: except for the very first vertebrates, all toothless vertebrates had ancestors *with* teeth. So why lose them?

In the late spring of 2021, a team of researchers at the Florida Museum of Natural History published a paper in which they concluded that frogs had a much higher rate of edentulism than any other vertebrate group. In fact, Daniel Paluh, the study's first author, told *Slate* that they "have lost their teeth more times than all other vertebrate groups combined."

So why frogs?

Since we've already seen how the first amphibians evolved from a small group of lobe-finned fish, it seems intuitive and/or only fair to start any survey of edentate vertebrates within the finny set. Although the detour turned out to be a short one, it did offer clues as to why 134 out of 429 frog species examined by Paluh and his team had lost their teeth.

To begin with, except for one bizarre occurrence, there are no reports of chondrichthyans (sharks and their relatives), extinct or extant, without teeth. That single recorded exception was a blackmouth catshark (*Galeus melastomus*) caught in a fishing boat's trawl off the southern coast of Sardinia in 2016. The specimen apparently lacked both teeth *and* skin but was otherwise reported to be "in good health and well developed." The baffled researchers speculated that the odd condition may have resulted from contaminated water, ocean warming, or "a genetic hiccup."

Until and unless additional specimens are discovered, I favor the latter explanation. And though it was only a single occurrence, the absence of both teeth and skin in this oddball shark provides an interesting reminder of the close relationship between skin and teeth in these creatures.

Among the bony fish, there appears to be only one order and two families where there is complete tooth loss: the order Gonorynchiformes (the milkfish), the Syngnathidae (pipefish and seahorses), and the Gyrinocheilidae (the ubiquitous aquarium denizens known as algae eaters).

The Gonorynchiformes are mostly toothless, with notably small mouths. Because of its commercial value in Taiwan and the Philippians, the milkfish (*Chanos chanos*) is probably the best known. Like the other members of the order, its diet consists primarily of tiny invertebrates, phytoplankton, and algae.

Masters of camouflage, the roughly fifty species of seahorses (genus

Hippocampus) and their slender, long-bodied relatives the pipefish (around two hundred species) inhabit coastal waters, often near sea-grass beds or coral reefs. Instead of scales, their bodies are encased in flexible bony plates. Consequently, they swim slowly—seahorses maintaining an upright position; pipefish with their bodies positioned horizontally. All species are toothless, and they sport long snoutlike structures that terminate at small upwardly tilted mouths. They feed on minuscule invertebrates, like copepods or shrimp, and they do so by making a slow, well-camouflaged approach, until the unsuspecting prey is located adjacent to their tiny, closed mouths.

What happens next is unique to fish that employ suction feeding, a technique that rests on the ability of the fish to rapidly expand the volume of its buccal (oral) cavity while keeping its mouth closed and temporarily sealing the rear exit to that chamber. This causes a drop in pressure within the buccal cavity relative to the surrounding water. Once the mouth is situated near the prey, it opens and the water just outside the mouth rushes in (moving from high pressure to low), carrying the miniature prey along with it into the mouth.

In the gyrinocheilids, teeth have been replaced by a sucker, situated around the downward-facing mouth. The structure enables them to latch on to smooth surfaces, where, true to their common name, they feed on algae. A less popular characteristic, but still significant to tropical aquarium enthusiasts, is the algae eater's fondness for consuming the protective barrier of slime found coating the bodies of most fish—the removal of which can lead to increased susceptibility to parasites and diseases. As an aside, this also explains why fishermen of any age should handle any fish they catch as little as possible during "catch-and-release" efforts.

The class Amphibia contains three extant orders, the Anura (also known as Salientia), which consists of 7,627 species of frogs and toads; the Caudata, which contains 805 species of salamanders and newts; and the Gymnophiona, made up of 221 species of legless, and admittedly weird, caecilians, whose consumption of their mother's skin (either before or after they're born or hatched, respectively) earns them a huge shout-out from fans of matriphagy—the consumption of the mother by her offspring.*

To keep a short story short, I could find no examples of edentulism in salamanders, newts, or caecilians, though there is certainly some variation among these taxa regarding the presence or absence of specific teeth on the various jawbones.

I spoke to Daniel Paluh, now an assistant professor at the University of Dayton, about his recent study on frogs. In it, he proposed that over the past two hundred million years, edentulism had occurred on more than twenty separate occasions. In a related finding, French researcher Tiphaine Davit-Béal and her colleagues reported that not one of the more than 350 extant species in the toad family Bufonidae has teeth. I asked Paluh why that was so.

One possible reason, he told me, is that most frogs (and bufonid toads) have evolved a highly specialized projectile tongue, which likely weakened the importance of teeth during prey capture. To envision the unique tongues in question, imagine yourself sitting at the breakfast table in front of a bowl of Cheerios. Everything looks normal, but inside your mouth, instead of your tongue being rooted in the rear half of your oral cavity, the base of this muscular structure sits just behind your lower incisors and canines. The remainder of the tongue extends backward, with the tip pointing away from the front of your mouth.

Okay, time to eat! But instead of sticking your tongue out a bit or flattening it down to accept a spoonful of cereal, you slowly move your head closer to the bowl. Then, as your mouth drops open, the body and tip of

* Species numbers are maintained daily by AmphibiaWeb, https://amphibiaweb.org/amphibian/speciesnums.html; these are from June 15, 2023.

the tongue flip forward at high speed, and the previously upward-facing surface slams down on the cereal bowl and its contents. Just as fast, the tongue snaps back into your gaping mouth, depositing some Cheerios at the back of your oral cavity, where they're swiftly swallowed. Only now do the jaws close.

All right, so maybe you shouldn't have envisioned this, but hopefully you get the concept here, since nowhere were teeth or chewing mentioned.

But why, I wondered, had edentulism arisen so many times in frogs? Paluh believes that toothlessness in this group was likely due to a shift—actually a delay—in tooth development. In most vertebrates, teeth begin forming during early embryonic development. But not in toothed frogs, where tooth formation is delayed until metamorphosis—the multistep transition from larval tadpole to juvenile frog. Though tadpoles do not have teeth, their tiny mouths are well equipped for algae scraping, something they accomplish with the aid of sharp-edged plates of keratin, sometimes fashioned into a beak. During metamorphosis, the tadpole's mini mouth undergoes a dramatic restructuring into a supersized frog mouth. The keratinized plates degenerate, replaced by tiny peg-like teeth.

Paluh suggests that late-forming structures may be more likely lost than early-forming structures. The key reason is that mutations or changes that occur during early stages of development are apt to be fatal, since they can disrupt the development of critically important structures like organs. The idea being that, if organs have already been established in an earlier developmental stage, mutations that occur *later*—as in the late-developing frog teeth—would have less chance of being harmful or lethal to the individual with the mutation. The most important organs are there already.

Paluh also brought up another interesting angle to the toothless-frog story, involving a phenomenon known as heterochrony, which is basically a change in developmental timing. If, for example, the development of an organism is truncated (i.e., has its duration shortened), early-forming traits are retained and late-forming structures never appear. This can lead to a type of heterochrony known as paedomorphosis (or neoteny), in which typically juvenile characteristics are retained by adults.

The classic example of the mudpuppy (*Necturus*), a giant salamander, should clear up any confusion. In most amphibian species, gills are lost as the larvae metamorphose into terrestrial adults. But *Necturus*, an oft-dissected specimen in comparative anatomy labs, retains these respiratory structures throughout adulthood. Scientists believe that this oddity may have occurred when a mutation resulted in a truncated developmental period, leaving the adult salamanders with larval gills. The fact that these individuals were fully capable of reproduction would have allowed the mutation to be passed on to future generations.

But why, you might ask, would gills in an adult salamander have adaptive value? One hypothesis is that a change in the terrestrial environment, like the arrival of a new predator, made it safer for salamanders to extend the aquatic larval stage. Those adult individuals that retained their gills would therefore have a selective advantage (i.e., be more likely to survive and reproduce) over those that no longer had gills and would have to go toe to toe with the predator. Here again, it was only when something in the environment changed that evolution occurred, since without the appearance of a new predator there would have been no adaptive value in retaining gills as an adult.

Similar mutations could have left adult frogs with the toothless condition previously reserved for tadpoles. As for environmental changes that might have favored toothlessness, it should be noted that the explosion in frog diversity that took place approximately two hundred million years ago was mirrored by similar increases in the diversity and success of two very important prey items: ants and termites. As we'll soon see, teeth are unnecessary for predators that have adopted a myrmecophagous diet (i.e., a diet consisting of ants and termites). In the case of some frogs and toads, this may have allowed toothless individuals to stay one hop ahead of their toothed competitors as they exploited a new food source.

Considering how successful ants and termites had become since the time of the dinosaurs, I wondered why hundreds of frog species still had teeth.

"Most frogs are considered generalist predators," Daniel Paluh told me, "and will attempt to eat anything, big or small, that can fit within their mouths. When a large prey item is targeted, a frog will typically use its projectile, sticky tongue to make first contact with the prey, but will then grasp and capture it with the jaws. When a small prey item is targeted, a frog will only use its tongue to capture the prey, completely bypassing the jaws. I think teeth likely still play a role when relatively large prey items are captured, and so it makes sense that they have been maintained in most frogs."

[12]

Dinosaurs, Turtles, Birds, and Dresser Drawers

A water snake glided smoothly up the pool, twisting its periscope head from side to side; and it swam the length of the pool and came to the legs of a motionless heron that stood in the shallows. A silent head and beak lanced down and plucked it out by the head, and the beak swallowed the little snake while its tail waved frantically.

—John Steinbeck, *Of Mice and Men*

For researchers currently in fields like vertebrate zoology and paleontology or those who work as taxonomists seeking to place reptiles (whether extinct or extant) into groups based on their evolutionary relatedness, the word "reptile" is problematic. As we know, humans love to sort things into groups, which is not necessarily a bad thing, since it often makes an otherwise unruly mess easier to deal with.

See, for example, your bedroom dresser: let's say that any object categorized as an article of clothing gets placed into the dresser. Similarly, linens would go into the linen closet, dry goods in the pantry, etc. There are separate drawers in your dresser for different items of clothing, like the things someone has previously named "socks," "underwear," "shirts," and "pants." Now, envision, if you will, that within each of those drawers there are dividers. So short pants go on one side of the divider, and long pants go on the other (see figure 28). The long pants section might be further divided into jeans and nonjeans sections.

"But what about my jackets and dress pants, Schutt?"

And, yes, this is becoming a very big dresser—but never mind that!

The naming of the items of clothing and their placement within their specific drawers and then in subdivisions of those drawers would one day become known as the science of taxonomy.* The groups of things in those drawers would become taxa (singular = taxon).

* Relating this model to real-life classifications, dresser = kingdom, each drawer = a phylum, each divided drawer section = a class, each subdivision of those sections = an order. Further subdivisions of those previous subdivisions would give you, in sequence: family, genus, and species—as well as a very strange set of drawers.

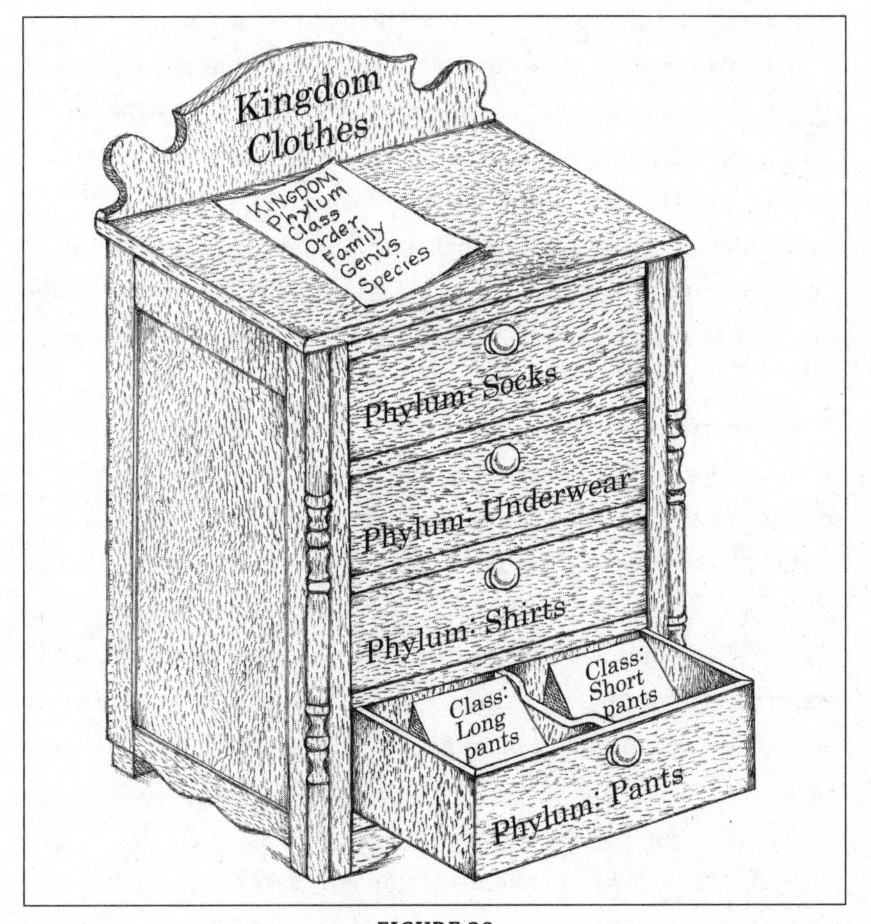

FIGURE 28

Beginning with the ancient Greeks, naturalists began placing the plants and animals they were encountering into their own drawers. Initially, the folks doing the sorting followed the dresser-drawer formula (which was kind of amazing, considering that this furniture hadn't been invented yet). In other words, they sorted out which creature or plant went into what group based solely on how similar those organisms were. Aristotle (384–322 BCE), arguably the first taxonomist, classified animals by the presence or absence of blood, and whether they lived in water or on land. Dolphins and sharks might have been classified together, since they both have the same general shape and live in water, and so might birds, bats, and butterflies, since they all have wings and fly.

As late as the eighteenth century, turtles, snakes, lizards, frogs, toads, salamanders, and the weirdo caecilians were all lumped together in the class Amphibia, and it wasn't until the nineteenth century that the four familiar classes of terrestrial vertebrates (Amphibia, Reptilia, Aves, and Mammalia) came into use. The anglicized versions of the four terrestrial vertebrate classes (amphibians, reptiles, birds, and mammals) eventually became, and still are, extremely popular with scientists and nonscientists alike.

As more and more animals and plants were named and categorized, the criteria for these designations moved away from mainly physical similarities, concentrating instead on evidence of evolutionary relatedness. This is where terms like "common ancestor" became important, since if species A and B shared a more recent common ancestor than either did with species C, then A and B were thought to be more closely related to each other. So bats and dolphins are more closely related to each other (since they share an ancient mammal ancestor) than they are to birds or sharks, respectively. Please note: If this sounds like a piece of cake, things can quickly become *far* more complicated.

As for what followed, to put this as painlessly as possible, beginning in the middle of the twentieth century, founders of the newly minted field of phylogenetic taxonomy decided (correctly) that for taxonomic groups to better reflect evolutionary relationships, each taxon (i.e., kingdom, phylum, class, order, genus, and species) would now be required to include *all* the descendants of a particular ancestor. In other words, in addition to socks, the sock drawer would now also contain stockings, pantyhose, and whatever else people down through the ages used to wrap around their feet.

The overarching term for this phylogenetic method became known as "cladistics," with valid taxonomic groupings referred to as "clades." If this seems difficult, I can assure you that it is, especially for those of us not well-versed in the terminology and the complex methods of data analysis. And things can get even more challenging, since modern taxonomists often deal almost exclusively with molecular traits like genetic sequences (which can be generated in great numbers), rather than physical characteristics like teeth (which can't). As a result, while the justification for employing cladistics remains sound, the accessibility of those concepts for nontaxonomists—and especially the public—frequently remains vexingly out of reach.

Meanwhile, modern paleontologists—initially, John Ostrom at Yale's Peabody Museum, in the early 1970s, and later, Mark Norell at the American Museum of Natural History—began gathering new evidence for an old hypothesis, namely that birds were an extant group of theropod dinosaurs, the likes of which include *T. rex* and the velociraptors. The problem was that dinosaurs had been included in the class Reptilia but birds hadn't been (having been assigned their own class: Aves). Since the dinosaurian ancestors of birds weren't included in the class Aves, modern taxonomists began thumbing their noses at the long-accepted classes Reptilia and Aves, which, as they explained, were not valid clades.

Today, new phylogenetically correct names have been developed to fix problems like the ones just described. For example, Sauropsida is now used to designate the extremely large clade containing the animals formerly known as reptiles (dinosaurs, lizards, crocodiles, snakes, and turtles) *and* birds, plus their common ancestors. Many scientists, though, continue to use "reptiles" when dealing with nonscientists, quite possibly to prevent rioting outside their zoo's Sauropsida House.

Okay, so about those toothless reptiles and birds . . . I mean, sauropsids.

Let's start with the easy stuff. There are no known species of toothless snakes or lizards.

In the birds and their ancient therapod dinosaur relatives (a clade within the Sauropsida known as the Avialae), tooth loss is thought to have occurred four times, including in the lineage leading to and including modern birds.*

The other sauropsids that went the edentate route were the turtles (through a separate event in an early turtle ancestor) and several groups of dinosaurs unrelated to birds. These included the nontherapods *Ornithomimus* ("bird mimic"), *Gallimimus* ("chicken mimic"), and *Limusaurus* ("mud lizard"), which all apparently had teeth as juveniles but not as adults. This is notable since *Limusaurus* joins the egg-laying platypus (*Ornithorhynchus anatinus*) as one of the only vertebrates known to completely lose their teeth as they mature, and with no replacement. This phenomenon is known as ontogenetic edentulism. In other words, these mammals have teeth when they are younger but lose them as they become adults.

Instead of teeth, adult platypuses have two pairs of thick keratinized pads called ceratodontes on their jaws. These are used to grind up the annelid worms (i.e., earthworms), insect larvae, crayfish, and shrimp that make up the platypus diet.

---

* There is ongoing debate about which theropod was the ancestor of birds.

Fated to become the most famous fossil in the world, a sauropsid known as *Archaeopteryx lithographica* resembles at first glance a small bipedal dinosaur, equipped as it is with a mouthful of sharp-looking teeth. A closer look, though, reveals something else: feathers. *Archaeopteryx* was first discovered in 1861 in an area north of Munich, and there are currently around a dozen specimens, varying in quality and completeness (one fossil consists of a single feather). Long thought to be the first bird, though it was eventually dethroned, *A. lithographica* remains for many paleontologists the best existing example of a transitional fossil—a fossil showing the remains of a life-form exhibiting both the characteristics of its ancestor and newly evolved traits. In this case, *Archaeopteryx* is a perfect blend of dinosaur and avian features.

As for how some sauropsids (like all modern birds) lost their teeth, much of what we know derives from several interesting studies—most of them involving chickens. For example, while investigating a gene named *talpid*^{2}, involved in organ development in chickens, researchers Matthew Harris of the Max Planck Institute and John Fallon at the University of Wisconsin discovered that the presence of a mutated form of the gene led to the production of sharp cone-shaped teeth in a sixteen-day-old chick embryo. In normal chicks, another gene, *SHH* (which is actually short for sonic hedgehog) is expressed in an area of the jaw that does not result in the production of teeth. But the presence of a mutant version of *talpid*^{2} apparently caused *SHH* to be expressed in a place on the jaw just right for the growth of teeth. The researchers described these teeth as looking like those of a crocodilian.

No toothed chickens have been hatched yet, but the work of Harris and Fallon, plus similar experiments by others, has shown support for the hypothesis that chickens maintain the genes responsible for tooth formation. These genes appear to have been turned off, however, as a result of genetic mutations that occurred more than one hundred million years ago. Although the timeline is different, this is essentially the same hypothesis that Daniel Paluh and his colleagues proposed to explain the twenty separate instances of toothlessness in frogs.

And just as some frogs were able to survive the loss of teeth because of the evolution of highly modified ant-snagging tongues, researchers believe that there was a different structure evolving at the very same time that tooth loss was taking place in turtles and other sauropsids. That structure was a beak—which clearly evolved on several occasions. Beaks (or rhamphothecas) are modifications of sauropsid snouts, in which the underlying jawbones are enclosed by sheaths of keratin. With circular gaps for the external nares (the openings through which many vertebrates breathe), these horny coverings are believed to have evolved from scales.

One hypothesis is that the first beaks covered only the tip of the jaw. Then, gradually enlarging and expanding posteriorly into the mouth, these structures eventually covered the socket-like alveoli, from which teeth normally grow. Once the alveoli were sealed off, the growth and replacement of the remaining teeth could no longer occur. Researcher and *Nature* paper coauthor Chun Li studied the ancient turtle *Eorhynchochelys sinensis*, and his findings support this hypothesis (see figure 29). In 2018, he told Agence France-Presse that this turtle, which dated from around 228 mya, had "a half-beak, half-toothed jaw," which he described as "an excellent transitional characteristic."

While many beaked creatures evolved plant-based diets (which included seeds and nectar), beaks also permitted some species to continue their carnivorous lifestyles in the absence of teeth. Beaks exhibit a wide assortment of shapes and sizes, enabling their use as tools, to pick, probe, tear, and crush. Beaks also help their owners manipulate the environment (to build nests, for example) as well as groom and preen (i.e., straighten and clean their feathers). Some birds even use their beaks to

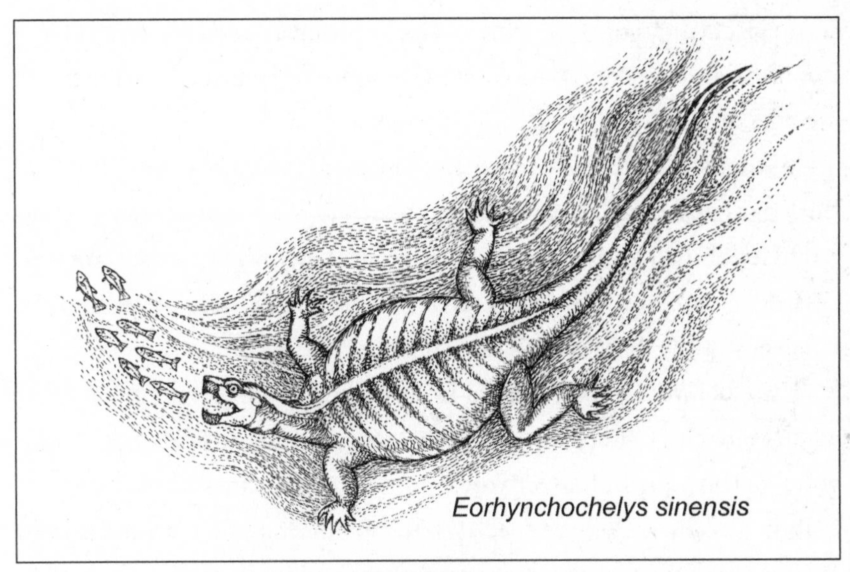

FIGURE 29

hold objects. Prime examples are two species of Galapagos finches (*Certhidea olivacea* and *Certhidea fusca*) that pick up cactus spines and twigs, wielding them like tiny crowbars to pry insects out of rotten wood.

Beak variation has also been important to researchers looking at how natural selection works. Peter and Rosemary Grant, Galápagos researchers extraordinaire, spent decades studying Galapagos finches, and famously proved that natural selection could lead to rapid changes in the beak size of the medium ground finch (*Geospiza fortis*).

Though tremendous variation exists in both beak form and function, most beak-bearing hunters fall under the heading of gape-limited predators. This is a term used to describe animals whose prey size is limited to what they can fit in their mouth whole. In birds, part of the reason for this is that their forelimbs have been modified into wings and are, therefore, incapable of assisting in actions like tearing apart prey. Raptors like hawks and eagles have gotten around this problem by evolving beaks that can tear prey into smaller chunks. The sharpened, sometimes serrated cutting edges of their beaks are known as tomia (singular = tomium), and the condition is also seen in turtle beaks, where the tomia are adapted for tearing vegetation as well as flesh.

Without the ability to chew their food, birds (and the beaked nonavian dinosaurs of the past) would seem to be at a digestive disadvantage compared to toothed creatures. As you'll recall, chewing is an important process that increases the surface area of food items, thus increasing the efficiency of the digestive juices it contacts.

Long before the fossil record of birdlike theropods became as clear as it is currently, there was the knowledge that many birds (i.e., granivores, herbivores, and insectivores) possess a structure that negates their inability to chew. In these species, after spending time in an esophageal storage bag known as the crop, swallowed food items, many whole or in relatively large pieces, pass into an initial region of the stomach called the proventriculus (which is narrow and lined with digestive glands) before moving on to the muscular-walled ventriculus, also known as the gizzard. This structure contains small stones and grit that the bird has swallowed. The contraction of the smooth, involuntary muscles that make up its walls turns the ventriculus itself into a gastric grinding mill, effectively transferring the job of chewing from the toothless mouth to the grit-filled gizzard. (This is also the reason some pet owners sometimes offer small amounts of grit to their caged birds every two years or so.)

As the dinosaur/bird link eventually became widely accepted by the scientific community, the function of the strange assemblages of smooth stones found in the abdominal region of certain theropod dinosaur fossils began to make sense. Christened gastroliths ("stomach rocks"), these stones have been found in several beaked dinosaurs belonging to the evolutionary line that would eventually lead to modern birds. In yet another example of evolutionary convergence, gastroliths were also present in the well-preserved body cavities of *Psittacosaurus*, a parrot-faced ornithischian dinosaur unrelated to birds.

This story does have its controversy, though—in this case related to the claims of some early-twentieth-century paleontologists that the giant herbivores known as sauropods (a group featuring *Apatosaurus*, the dinosaur formerly known as brontosaurus) also had gizzard-like grinding mills said to contain gastroliths. These long-necked behemoths clearly required a massive daily intake of plant matter. Since their small teeth seemed woefully inadequate for the job of processing such tonnage, it was hypothesized that these giants also evolved the equivalent of gizzards containing gastroliths. Though the idea of giant dinosaurs with gizzards is now rather widespread, it has been challenged by other paleontologists, including a pair of German researchers, who discount what they consider scant proof of sauropod gastric mills and/or gastroliths. The latter, they claim, are extremely rare, though given the relatively extensive fossil record for the sauropods, they shouldn't be. These paleontologists hypothesize that rather than gastric mills responsible for smashing up plant matter (and leaving behind stony evidence in the fossil record), the process of plant digestion in sauropod dinosaurs took place via a method far more common across the animal kingdom. They suggest that, aided by endosymbiotic bacteria, sauropod digestive systems resembled huge fermentation vats, processing food slowly and ultimately leaving no trace in the fossil record.

Sticking to the topic of puzzling paleontological questions, although *Archaeopteryx* stands as a wonderful exception, the scarcity of transitional fossils remains problematic for many nonscientists. For example, we know that bats evolved from nonflying ancestors, but there are currently no known fossils to inform us about what the ancestors of the first bat were doing right before they exhibited true flight. Were they gliders, perhaps something along the lines of flying squirrels? That seems likely.

But whether the absence of protobat fossils is due to the overall rarity of delicately boned creatures like bats in the fossil record, or the likelihood that true flight evolved rather quickly (leaving few fossils), we do not have a transitional fossil to clue us in.

What we do have, however, is evidence that, although they don't have them now, birds originally had teeth. One such critter was *Hesperornis regalis*, a flightless cormorant-like diving bird that lived during the Cretaceous period (around 100–65 mya). (See figure 30.) First discovered in the early 1870s, *H. regalis* is both well represented and well preserved in the fossil record. What makes it even more spectacular is that it exhibited both teeth *and* a beak, a convergent trait also exhibited by ceratopsian dinosaurs like triceratops. Far older and far better known is the previously mentioned toothed but beakless bird *Archaeopteryx lithographica*, from the Late Jurassic period (roughly 150 mya).

These fossils and others further solidified the relationship between dinosaurs and birds—a connection that had its origin in the nineteenth century. As far back as 1877, paleontologist Othniel Charles Marsh wrote, "It is now generally admitted by biologists . . . that birds have come down to us through the Dinosaurs." And a decade later, Darwin's bulldog, Thomas H. Huxley wrote: "Surely there is nothing very wild or

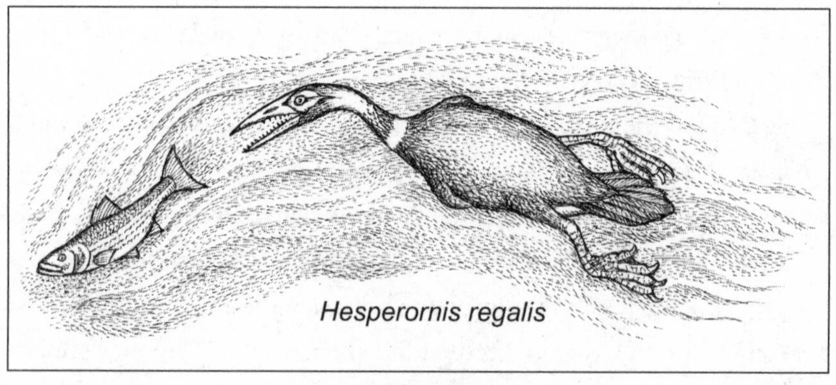

FIGURE 30

illegitimate in the hypothesis that the phylum of the class Aves has its foot in the dinosaurian reptiles."

A century after that, paleontologists like John Ostrom, Mark Norell, and their colleagues put an exclamation point on the dinosaur/bird relationship. Having gathered an abundance of new evidence to support an old hypothesis, their research served to illustrate that the real strength that characterizes every field of science can often be seen as a lengthy—sometimes extremely lengthy—accumulation of evidence.

[13]

Toothless Mammals, from Anteaters to Whales

I went to the zoo once and saw this thing they call an anteater.
That was quite enough for me.

—Often attributed to Thomas Pynchon

Although the vast majority of mammals have teeth, within the class Mammalia, toothlessness has apparently evolved four separate times. But here there are no beaks to speak of. Instead, the success of the approximately twenty-seven species of toothless mammals (out of approximately six thousand species) was clearly dependent on the evolution of alternative feeding structures and the often-specialized feeding behavior these structures made possible.

The order Monotremata is a small group of egg-laying mammals, which includes the duck-billed platypus (which has teeth) and four species of toothless echidnas. With their dome-shaped bodies, covered in variations of porcupine-like spines, echidnas roughly resemble hedgehogs—if the hedgehogs in question were breathing and feeding through a long tube. But more on that later. Monotremes are thought to have been an early branch on the mammal tree that sprang from the therapsids, a group of extinct vertebrates that were the dominant land vertebrates throughout the Permian period (299–251 mya). Early members of this diverse order included the sail-backed dinosaur-toy-collection impostor

Dimetrodon, while later groups began exhibiting what would be defined a quarter of a billion years later as "mammal-like" characteristics.

What is also clear is that only a handful of therapsid species survived the massive Permian-Triassic extinction that occurred around 251.9 mya. As these global extinction events tend to go, once the dust settled, there were plenty of open niches for the survivors to fill. In this case, the survivors included two small groups: the dinosaurs and the therapsids. Among the latter group were the dicynodonts (discussed earlier because of their tusks) and the cynodonts—who first appeared in the Late Permian period and diversified after the Permian-Triassic hammer had been lifted.

Part of that cynodont diversity was a clade that became known as the Mammaliaformes ("mammal-shaped"), which some researchers consider to be the first mammals. From the fossil record, we believe that they began appearing in the Late Triassic (roughly 210 mya), looking very much like modern-day shrews,* which, admittedly, was not a great size to be if you're looking to leave an extensive fossil record of your existence. As it turned out, it *was* a good thing if you were trying to eke out a living in the shadow of the dinosaurs, which had been running the show from just after the Permian-Triassic extinction until they were pretty much wiped out 65 mya, leaving only the birds to mess up everyone's dresser drawers.

Most of this interesting stuff took place around 250 million years ago, so the evidence of mammalian ancestry rooted among the therapsids relies primarily on characteristics preserved in skeletal structures like skulls, teeth, and limbs. The fossil record also provides indirect evidence of nonbony traits, like the presence of hair, mammary glands, and even increased metabolic rates and endothermy (i.e.,

* *Tikitherium*, identified from a single upper left molar found in India in 2005, may push back mammal origins to somewhere closer to 225 mya.

warm-bloodedness). Because of the totality of evidence, and with nothing to counter it, there is currently little argument about the therapsid ancestry of mammals.

Now that we know where mammals came from, let's get back to the monotremes, specifically the echidnas. The key reason that the monotremes have been placed near the base of the mammalian tree is that they share a number of "reptilian" characteristics with their ancestors: most famously, the fact that they lay eggs.* Along with its fellow monotreme, the platypus (which, as was recently mentioned, has teeth but loses them by the time it waddles out of its breeding nest), the toothless echidnas did not diverge into myriad species, like the rodents and bats did. Two of the four monotreme families went extinct, outcompeted by the therians (marsupials and placental mammals), a lineage that evolved live birth and a slew of not-insubstantial adaptations like larger brains, more efficient red blood cells (able to carry more oxygen), and teeth adapted for nearly every conceivable diet. It is likely that the echidnas persisted because, like the edentate anteaters and pangolins, which are both therian mammals, they convergently evolved specialized versions of a structure that took the place of teeth. That structure was their tongue.

I spoke to my grad-school friend Karen Reiss, now a professor of biology at College of the Redwoods and a leading expert on the structure and evolution of tongues in anteaters, pangolins, and echidnas. I asked her to give me the lowdown on her toothless study animals.

Except for the baleen whales, she told me, "Edentulous mammals are for the most part ant and termite eaters, myrmecophages."

"In other words," I replied, "they're dietary specialists."

---

* Like squamates (e.g., snakes and lizards), echidnas and duck-billed platypuses possess a single egg tooth (or caruncle) at the end of their upper jaw that they use to free themselves from the confines of their shell. It is lost soon after they hatch. In squamates, this structure is a real tooth, though in egg-laying mammals it probably isn't—since it lacks enamel.

"Precisely," she said. "And they're all known for their extraordinarily long, sticky, fast tongues . . . And their tongues are indeed magnificent. I was and still am pretty blown away by how similar, on a really detailed anatomical level, the tongues of these unrelated taxa are. It's an extraordinary example of convergence."

Though they lack an ant-eating common ancestor, echidnas (from Australia and New Guinea), pangolins (from Asia and sub-Saharan Africa), and anteaters (from South and Central America) all exhibit similar modifications of their head musculature. These anatomical tweaks have resulted in a long, cylindrical tongue detached from the hyoid apparatus, the bony structure that usually anchors the tongue to the initial section of the trachea. Instead, these tongues are attached along the sternum (or breastbone), which extends back toward the hips.

The extreme example of this phenomenon can be found in the eight species of pangolins (see figure 31), bizarre-looking creatures currently under siege because of their huge, overlapping keratinous scales, which are used in traditional Chinese medicine. Below its formidable-looking body armor, the pangolin's sternum curves back like a bony divining rod, until it finally terminates near the pelvis. This section of the sternum also serves as the attachment site for the base of the tongue, which can extend nearly a full body length beyond the elongate snout it shares with the anteater and the echidna.

"Detachment from the hyoid means that the tongue is able to move more freely fore and aft," Reiss said. "And a long, thin cylinder of muscle can get much longer quickly by just narrowing its diameter a bit."

Oddly, what she was describing reminded me of the movement of earthworms, which have bands of muscles wrapping around their bodies while other muscles run from front to back. Contraction of the wraparound muscles shoots their anterior ends forward (kind of like when you squeeze the middle of a long water balloon). Then, after gripping the ground with tiny anteriorly located bristles called setae, the muscles that run along the length of the earthworm's body contract, pulling the rest of the creature forward.

Reiss affirmed that this was indeed reminiscent of the way her termite-and-ant-munching study subjects ejected and retracted their muscular tongues, which, she pointed out, are also oval-shaped in cross section, like worms. Examples of convergence between the pangolin, the anteater, and the echidna can be found in the large claws that each species uses to tear open an ant or termite colony before shooting its megatongue into the twists and turns of the nest's interior.

"When the tongue is pulled back and then shot out again, there are usually bristles or ridges on the palate that help scrape the prey off before the next incursion," my friend told me. This is an efficient way to ingest the largest number of insects in the least amount of time, which is important from both a nutritional and energy standpoint. It also enables these animals to avoid what Reiss termed "some gnarly defenses" exhibited by the ants and termites. Fortunately, the tiny insects' bitey

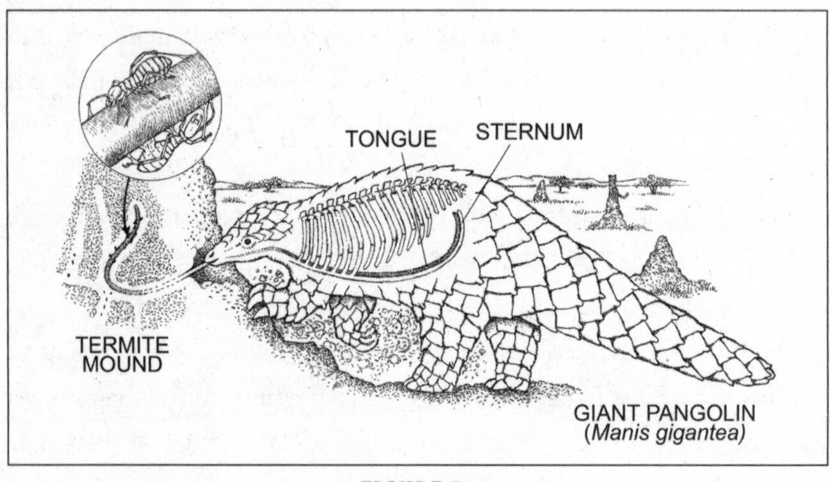

FIGURE 31

counterattacks take a minute or so to get mobilized, and by then the toothless myrmecophage has likely moved on to assault a new nest.

Continuing the theme of convergence, Reiss noted that pangolins have a gizzard-like region of their stomachs, containing gastrolith-like stones.

"Un-friggin'-real," I responded, stepping into serious science-speak mode. "But it still doesn't explain how these strikingly similar tongues showed up in the three unrelated ant-eating mammals."

"True," she said, "but it suggests that on a developmental genetics level, there are only a few tweaking points in the development of the head that can transform the preexisting mammalian chassis into something that has a long and highly extensible tongue."

"In other words, there aren't many ways that mutations can produce this type of ant-snatching tongue?"

Reiss nodded.

"Very cool," I replied, and my old friend laughed, both of us marveling at yet another amazing aspect of nature and thinking how lucky we were to be studying it.

THE FOURTEEN LIVING species of baleen whales that belong to a clade known as Mysticeti are all toothless, but like their toothed whale relatives (the Odontoceti), they are thought to share a common ancestral group, the Archaeoceti ("ancient whales"). These prehistoric toothed whales, some of them equipped for an amphibious lifestyle, showed up around fifty million years ago and are themselves thought to be closely related to the anthracotheres, the skinny-hippo-looking ancestors of both modern hippopotamuses and the first whales.

Tiphaine Davit-Béal, senior lecturer at Sorbonne Université, summed up mysticetous toothless trends in her excellent paper on tooth loss. "In the ancestors of mysticetes the teeth became smaller, and the animals adapted to filter-feeding, which allows access to small prey. This change in feeding [habits] was facilitated by the simultaneous presence of rudimentary baleen plates." (Baleen is commonly known as "whalebone," but incorrectly, since it is actually composed of the structural protein keratin.) Tweaked by natural selection, as evidenced by the fossil record, these ancient baleen plates initially sat beside the teeth they would eventually replace.*

As baleen increased in size and complexity, its position in the mouth evolved as well. In later species (including some extant whales), it hung down from the palate in close-knit fibrous tubes, forming a pair of comblike structures on either side of the whale's mouth. These baleen plates effectively transformed their owners from tooth-bearing carnivores into toothless filter feeders. With decreased selection pressure to maintain them, teeth became smaller and smaller as baleen continued to evolve into an efficient tool—separating multitudes of tiny fish and small marine invertebrates, like krill and zooplankton, from the giant gulps of water that previously contained them.

Several variations on this basic feeding pattern can be seen in extant baleen whales—from plowing open-mouthed through the sediment for tiny bottom dwellers, to blowing rings of bubbles called bubble nets, which entrap the encircled fish before they're gulped down. Although baleen whales include the blue whale (*Balaenoptera musculus*), the largest animal on the planet, they follow a trend that spans all toothless mammals, feeding on prey that are minuscule in size and available in vast numbers.

The toothless vertebrates have carved out their own narrow dietary niches, but only after evolving toothlike adaptations like beaks or baleen.

---

* If this sounds familiar, recall that teeth and beaks appeared together in some ancient birds before teeth were eventually lost.

Given the importance of teeth to the vast majority of vertebrates, it should come as no surprise that teeth have been vitally important to the evolutionary success of humans as well. But they've also been key to success of a different kind, one that reflects the power, status, and health of the individual. And similar to what occurs in the animal kingdom, human success in the absence of teeth can present its own set of challenges. Perhaps James Brown, the Godfather of Soul, said it best: "Hair is the first thing. And teeth the second. Hair and teeth. A man got those two things he's got it all."

PART IV

Human Teeth:
The Bad Old Days and Beyond

[14]

A Man of Few Words . . . and Fewer Teeth

My master observing that he had no teeth in the fore part of his jaws, asked the boy by what accident he had lost them. "By no accident," replied the sweep, "my mother sold them when I was young, to a dentist, who transplanted them into the head of an old lady of quality."

—HELENUS SCOTT, *THE ADVENTURES OF A RUPEE*

WHEN I WAS in first or second grade (sometime just after Long Island had been dropped into place by a glacier), like millions of other American schoolchildren, I was taught the moralizing tale of how a youthful George Washington cut down a cherry tree and then, feeling guilty, immediately fessed up to the crime. Since our feelings for the adult Washington were of extreme reverence, this story fit right in with how most of us figured the Revolutionary War hero, Founding Father, and first president must have behaved as a kid. This is not to say, mind you, that my friends and I wouldn't have dragged off the slaughtered cherry tree, dug out the stump, and covered up all remaining evidence of its existence with leaves. But we did get the message: George Washington never told a lie.

Harder to explain was the story that often accompanied the cherry-tree confessional. It was, of course, the story of George Washington's wooden teeth. "It's why you never see a picture of him smiling," I remember my mom telling me. But while that explanation certainly made sense to me at the time, I don't remember it coming off as the sort of brush-your-teeth-or-else cautionary tale we might have expected in the sequel to "Naughty Boy with a Hatchet Fesses Up."

Much more recently, as I approached this enduring legend from a different perspective, it seemed to get even more puzzling. Why weren't Washington's dentures made of substances that had been commonly used at the time, like ivory or porcelain—or, for that matter, anything superhard and termite-proof? As it turned out, they were. Delving a bit deeper, it also became apparent to me that dentures back then were quite rare.

In 1991, impending construction at Christ Church outside London necessitated the removal of a roughly Georgian-era crypt located underneath. It was a perfect opportunity to disinter and examine the coffin-bound inhabitants before reburying them. Of the 968 bodies buried between 1729 and 1852, only 12 of them showed evidence of dental work and only 9 of these had dentures. Though there exists the possibility that some of the dental ware was removed before burial, it is more likely that its absence had more to do with the fact that dental appliances were expensive, and few people could afford them.

Today, of course, information revealed through commonly performed dental work such as fillings, caps, implants, root canals, and dentures form a kind of dental fingerprint, enabling forensic odontologists to identify deceased individuals who might otherwise be unidentifiable. One of the first reported examples of this occurred in March 1766, when Boston silversmith Paul Revere (who occasionally dabbled in dentistry) was asked to identify the badly decomposed body of Major General Joseph Warren, who had been killed by British soldiers on June 17, 1775, during the Battle of Bunker Hill. Since the colonists lost that battle, they were unable to recover Warren's body until the redcoats left Boston the following year.

If Major General Warren's name sounds familiar, it might be because on April 18, 1775, the Harvard-trained doctor had instructed Paul Revere to set off from Boston to warn the colonists in Lexington that the British were planning to arrest John Hancock and Samuel Adams. After that, Revere was supposed to hightail it to Concord to alert the rebels that their munition stores there were the next thing on the redcoat to-do list. According to historians, though, Paul Revere never uttered his famous cry "The British are coming!" In fact, he never made it to Concord. Instead, he and two other men were detained by British troops. Revere was relieved of the horse he had borrowed, then eventually released, after which he walked back to Lexington. Luckily, Samuel Prescott, who had also been detained but escaped, made it to Concord with the warning.*

Nearly a year later, Revere and several of Warren's relatives disinterred his corpse, which had lain in a shallow grave with another soldier's. At that point, Revere was able to perform what was likely the first forensic identification of an American serviceman based on his dental records, specifically, a dental bridge consisting of a pair of ivory teeth that Revere himself had constructed shortly before the battle, to replace the major general's upper left canine and first premolar.

During the mid- to late eighteenth century, those citizens who could afford artificial teeth, often found themselves wearing a faceful of heavy, unwieldy-looking contraptions. These were generally crafted by ivory turners and metalsmiths at the expense of walruses, elephants, and hippos. Less exotic teeth were also used, mainly from cows and horses. All four sets of George Washington's surviving dentures contain teeth from different branches of the class Mammalia. Two of these appliances, though, also contain something else: human teeth.†

---

* Revere got the fame, though, after the *Atlantic Monthly* published Henry Wadsworth Longfellow's "Paul Revere's Ride" in January 1861.

† According to historian Jennifer Van Horn, it is likely that "Washington commissioned six pairs of dentures over the course of his lifetime."

This information had come to light in the 1970s, thanks to a study undertaken by Reidar Sognnaes (1911–1984), a dental pathologist and college professor who would go on to become the founding dean of the UCLA School of Dentistry. In addition to his careful examination of Washington's dentures, Sognnaes had gained a measure of fame by comparing information on Adolf Hitler's antemortem and postmortem dental status. The latter consisted of a jawbone, some teeth, and a gold bridge recovered from the pit where the Russians had reportedly burned the bodies of Hitler and his newlywed bride Eva Braun. By comparing these charred remains to x-rays taken previously by Hitler's dentists, Sognnaes was able to quash rumors that the dictator had somehow escaped, confirming instead that Hitler had died by suicide in his besieged Berlin bunker on April 30, 1945.

In a paper published during the United States' bicentennial year, Sognnaes presented research on all four of Washington's surviving dentures, paying special attention to their composition, and going so far as to produce precise models of the dental relics. Sognnaes determined that in addition to animal and human teeth, the other materials used to fashion the Founding Father's dentures included "gold springs, gold pins, a swaged gold plate [i.e., shaped with a hammer], two bulky lead bases [that fit onto the upper and lower jaws], several iron rods, steel springs and six wooden fastening pegs [which, in one case, secured some of the teeth to their lead base but were not visible and so could not be considered wooden teeth]." (See figure 32A.)

According to Sognnaes, Washington's first set of dentures (or at least his oldest surviving set) was produced in 1789, the year of his inauguration, and the work was done by the renowned New York dentist John Greenwood. The prosthesis he built was a partial, and Greenwood, who has been called "America's first native-born dentist," took advantage of the fifty-seven-year-old Washington's single remaining tooth (a left lower premolar), fashioning the denture so that it fit over and was partially held in place by the lonely bicuspid. (See figure 32B.)

Sognnaes wrote: "Some of the other teeth which Greenwood inserted into this denture are also thought to have been from Washington's own mouth."

To explain the seemingly odd presence of human teeth in a denture, a brief trip through some of the relevant dental practices of the time will be helpful.

BY WASHINGTON'S LIFETIME, the extremely *un*scientific job of tooth pulling had been around in various cultures for over two thousand years. Well-documented in Europe beginning in the Middle Ages and lasting into the early nineteenth century, tooth pullers often set up shop in markets and fairs, making them seem more like performance artists than medical types. It is likely that this theatrical effect was enhanced by the

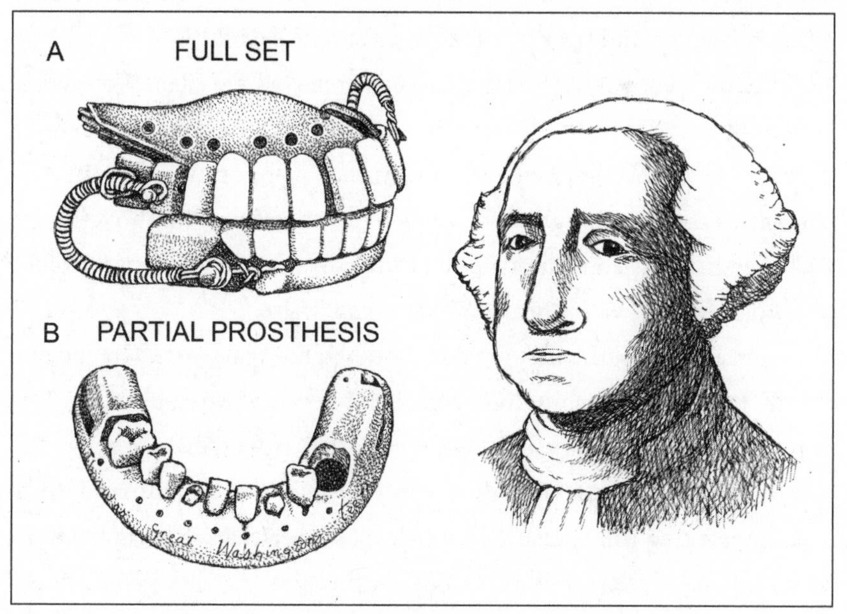

FIGURE 32

presence of assistants who played music or capered about in comical outfits, to drown out the screams of the unanesthetized and draw attention away from their considerable agony. Quackery was common, as were infections and significant numbers of tooth-related deaths.

A step above the tooth pullers were the so-called barber-surgeons, who made their first appearance in Europe during the Middle Ages. They were popular into the eighteenth century—though given the painful nature of their profession, "busy" might be a more appropriate description. Among a rather long list of duties, barber-surgeons cut hair, performed minor surgery, bled patients for a variety of ills (with or without medicinal leeches), and extracted teeth.

The jack-of-all-trades nature of the barber-surgeons underwent a transformation in France during the first half of the eighteenth century. Some of them began to train and specialize, abandoning pursuits like setting broken bones, lancing boils, and administering enemas. Instead, these *dentistes* devoted themselves full-time to filling, scraping plaque from, replacing, and transplanting their patient's *dents.*

The emergence of the first dentists (also referred to as dental surgeons) roughly coincided with a dramatic rise in the consumption of refined sugar—a topic that will make an unfortunate reappearance shortly. And though George Washington was reportedly not a sweet-tooth victim, he did have his first adult tooth pulled in 1756, when he was twenty-four years old. Dental-wise, it was downhill from there.

From 1789 until the first president's death ten years later, Greenwood sought to remedy Washington's considerable dental woes, building dentures for him and repairing them when they broke down (e.g., replacing the hinge springs). He pulled Washington's last remaining tooth in 1796, preserving this treasure in a tiny glass case that he hung from his watch chain. Greenwood also told Washington that his penchant for port wine was responsible for turning his dentures black, calling it "very pern[i]cious to the teeth." In a letter, he advised his patient to "take them out After dinner," place them "in cleain water" and "Cleain them with a brush and som Chalk scraped fine," which would "Absorbe the Acid which Collects from the mouth and preserve them longer."

With a new emphasis on both dental treatment and preventive care, some eighteenth-century practitioners of the newly minted field of dentistry began to steer their profession toward a more scientific path. One of the most influential was the French surgeon Pierre Fauchard (1678–1761), often called the Father of Modern Dentistry. Fauchard stressed the dangers of sugar consumption and the acids it produced—though he had no idea that the acids were being released by oral bacteria using the sugar as an energy source. He also advocated dental fillings of tin, lead, or gold to treat dental caries (cavities) and the use of dental prostheses in which artificial teeth were wired to healthy ones. Fauchard passionately denounced charlatans, who often put the lives of their unsuspecting patients in danger. He promoted the new concept of daily tooth care, such as using toothbrushes, though he warned against the inclusion of abrasives like ground-up bricks and pumice stone as dentifrice ingredients.

Fauchard might have also won the title of the Father of Not Quite Modern Dentistry with a recommendation that his patients use their own urine as mouthwash. Shockingly, many people stuck to alternatives like toothpicks (sticks, quills, and feathers were popular), tooth powders and toothpastes—some of these containing ground-up mollusk shells—treatments that had been around in various forms since Rome in the first century CE.*

By the second half of the eighteenth century, French and English dentists had become popular, and several of them moved their practices from Europe to the American colonies. Here, they became highly

---

* CE ("of the Common Era") and BCE ("before the Common Era") are used throughout.

successful—often taking up temporary and well-publicized residencies in cities along the East Coast. For George Washington, the most important of these imports was the French dentist Jean-Pierre Le Mayeur, who left British-occupied New York City in 1781 and headed south.* Apparently, Le Mayeur was intent on finding some mouths that weren't quite as critical about the Franco-American alliance as those belonging to the British officers he had been treating in New York. He met Washington soon after, becoming his dentist and family friend, and treating him through the 1780s. And although the Founding Father was ministered to by a half dozen or more dentists over his lifetime, it was Washington's regrettably underdocumented association with Le Mayeur that would lead to a serious controversy some 235 years later.

But first things first.

By the late eighteenth century, it was common practice for well-to-do patients to collect their own dwindling supply of teeth after they were either yanked out or fell out of their own accord. Rather than saving them for the tooth fairy (who didn't begin working until 1908), teeth were sometimes kept so that they could be incorporated into custom-fit dentures. In a 1782 letter, Washington described the location of his own stash. "In a drawer, in the Locker of the Desk which stands in my Study, you will find two small (fore) teeth; which I beg of you to wrap up carefully, & send inclosed in your next Letter to me. I am positive I left them there, or in the secret drawer in the locker of the same desk."

As for who else might have had a hand (or two) in George Washington's mouth, silversmith Paul Revere (1735–1818) was reported to have pulled some teeth in his time, but his skills were far better suited for crafting dentures like the ones he made for Major General Warren. To be filed under Other Stuff Paul Revere Never Did, and despite long-standing

* Le Mayeur was also known as Le Moyer, Le Mayeaur, and Joseph Lemare.

rumors to the contrary, there is no evidence that he ever worked on dentures belonging to George Washington.

In addition to obtaining teeth from their patients, dentists like Le Mayeur, and later Greenwood, acquired them through advertisements directed at those who wished to make money by selling teeth right out of their own mouths. In Victor Hugo's *Les Misérables*, the doomed character Fantine endures the gruesome practice, sacrificing her two upper incisors to raise money to save her child. "The candle illuminated her countenance," Hugo wrote. "It was a bloody smile. A reddish saliva soiled the corners of her lips, and she had a black hole in her mouth. The two teeth had been extracted."

In real life, Fantine's teeth would have been pulled without anesthesia, either for use in someone else's dentures or as donor teeth in the admittedly strange practice of dental transplantation—a procedure thought to have originated in France around 1850. After purchasing teeth from a destitute donor or an unsavory character who might have pulled them from the bodies of the dead, surgeon dentists would insert the teeth, roots-first, into the recently vacated alveoli of a recipient patient's jaw.

Some three decades earlier, teeth obtained by unscrupulous methods became known as Waterloo teeth, after the Battle of Waterloo. That famous clash, between the forces of Napoléon Bonaparte and the Duke of Wellington, took place on June 18, 1815. In the aftermath, the bodies left on the battlefield were robbed of their belongings and worse. Many of the fallen also had their teeth yanked out and sold to dentists, who used them to create dentures. Similar desecrations were said to have occurred on the battlefields of the American Civil War. Other teeth bound for use in dentures or transplants included those collected from executed criminals and the disinterred dead.

I also found that beyond the improbable nature of tooth transplantation, the procedure was an example of how news did *not* travel fast back in the eighteenth century. In this and other cases, it sometimes took four or five decades for an old (and often ineffective) medical practice to be supplanted by a newer and better one. A contemporary of Le Mayeur wrote that during the Revolutionary War, Washington's soon-to-be dentist transplanted "over one hundred teeth" during the winter of 1781–1782 "and not one succeeded." Le Mayeur somehow neglected to include this information in the newspaper ads he took out, proclaiming instead the miracles of tooth transplantation and his own flawless record of success.

But Le Mayeur wasn't the only person promoting the procedure. In 1771, John Hunter (1728–1793), the Father of Scientific Surgery, authored *The Natural History of the Human Teeth*. In it, Hunter wrote about tooth development and anatomy, as well as several dental procedures, including tooth transplantation. Hunter was a medical heavyweight of his time, though he took his studies of venereal disease to the slightly more than extreme by infecting his own penis with gonorrhea-contaminated pus. This he did in an effort to acquire the disease for study purposes—reportedly achieving a bonus by contracting syphilis as well. Beyond Hunter's triumph in giving himself two STDs, he reportedly performed the first successful artificial insemination on a human.

Hunter also formalized the classification of human teeth, from front to back, as incisors, cuspids (canines), bicuspids (premolars) and molars. Note that in humans, cusps are not present on incisors, and canine teeth each bear a single cusp. Premolars are equipped with a pair of cusps (hence the term "bicuspid"), and there are four or five cusps on the occlusal surface of each molar.

As it turns out, Hunter was a strong proponent of tooth transplantation, and he attempted to support the scientific validity of the practice in a way that apparently made a great deal of sense to him back then—that is, by grafting a human tooth onto the comb of a rooster. (See figure 33.) He followed up by stitching a rooster testicle onto the belly of a hen.

With the feasibility of the tooth transplantation procedure now confirmed (at least in Hunter's mind), he emphasized to his readers: "It will scarcely be necessary to observe that the new Teeth should always be perfectly sound, and taken from a mouth which has the appearance of that of a person sound and healthy; *not that I believe it possible to transplant an infection of any kind from the circulating juices*" [emphasis mine].

Reading Hunter's writing on the cockscomb experiment, it becomes clear that the great surgeon may have had his own doubts about the procedure. "I may here just remark," he wrote, "that this experiment is not generally attended with success. I succeeded but once out of a great number of trials." With the gift of hindsight, this outcome is anything but a surprise.

Having lived what was an undeniably interesting life, Hunter died

FIGURE 33

from a heart attack in 1793, and soon after his death, some of his work appeared—though with a disturbing twist. His surgeon brother-in-law (and former student), Everard Home, began cranking out award-winning scientific papers at a prodigious rate. This surprising surge in productivity, and the acclaim that followed, led to Home being named vice president of the Royal Society. He also served as surgeon to King George III, was knighted, and was elected president of the Royal College of Surgeons. For years, the trustees of a museum dedicated to Hunter's work had been trying to wrest legal control of Hunter's copious collection of notebooks and unpublished papers away from his now hugely successful brother-in-law. But in 1823, instead of handing the material over, and for reasons that in retrospect do not defy logic, Everard Home burned them.

Hunter's preserved specimens survive, though, including the cockscomb, still attached to its owner's neatly bisected head. With its carefully halved and still-embedded human tooth, it currently resides at the Hunterian Museum in London.

As an anatomist, it was hard for me to imagine how the tiny arteries running from a tooth transplant recipient's jaw into their tooth, once severed by the removal of that tooth, could somehow grow up and into the roots of a newly inserted donor tooth. Further, having made that improbable journey through the minute apical foramen at the tooth root tip, this vessel would then need to connect to the severed capillary bed within the presumably living (more likely, formerly living) portion of the donor tooth, the so-called tooth pulp. After supplying nutrients and oxygen to that donor tooth, the blood carried within this newly reconnected vasculature would then need to exit the tooth via a similarly newly grown vein, for an eventual return trip back to the heart.

In other words: nope.

Puzzled by reports that some transplanted teeth could in fact last for a year or two, I brought up the topic with my own dentist, Kevin Sheren, a high-tech dental wiz practicing on Long Island's East End.

"These were dead teeth, right?" I asked him during a visit to remedy a minor dental problem I had been having.* "So how on earth could they have lasted for any length of time?"

"You're on the wrong track," Sheren said. "The transplanted teeth *would* be dead, but what they were looking for was for the recipient's jaw to ankylose around the tooth—basically, to fuse to it—very much like your jaw ankyloses around a modern implant."

The problem, he continued, was that nobody knew about infections or the fact that while the tooth might be initially held in place by soft tissue, the odds of a nonsterile object, imperfectly shaped and placed, finding a new and permanent home, were incredibly low.

Looking at the procedure from this new angle, the reports of temporary tooth-transplantation success now made some sense, so I followed up. "And since the transplanted tooth was no longer alive, it would probably crack after anything like normal use—if it ever got to the point where you could use it."

Sheren nodded. "That's right."

But even back in the eighteenth century, some medical professionals, including the famous French dentist and inventor Nicolas Dubois de Chémant, argued against the procedure. "Transplanted teeth can never recover life, as the public and even some practitioners were led to believe," he wrote. He added that it was a highly effective means by which one could "communicate the venereal virulency or the scrofulous infection," that is, venereal disease or tuberculosis of the lymph nodes. The Parisian escalated his disagreement with Hunter by stating that "the

* Because of this, my first question to him probably sounded more like "Ee ah ed eet ight?"

dangerous practice of transplanting human teeth ought to be for ever banished from the profession of a dentist, and they ought not to be made use of in any manner whatsoever."

By way of an alternative, in the 1804 edition of his work *A Dissertation on Artificial Teeth*, de Chémant said that he had "examined almost all the substances in the mineral kingdom, and at length composed a paste, which, when it is baked has every desirable advantage." The hardened final product turned out to be porcelain, which de Chémant used to craft dentures. Ever the entrepreneur, he promoted his mineral paste prosthetics in advertisements, on one occasion adding a somewhat-less-than-related plug for a dining table he had also invented.

Notwithstanding the stomp job de Chémant applied to tooth transplantation, the problematic procedure became popular in England and eventually the United States, hanging around until around the turn of the nineteenth century. By then, porcelain dentures had become more affordable and thus more commonplace.* It is also not a stretch to assume that folks grew tired of contracting syphilis from someone else's infected incisors or having their recently transplanted teeth fall into their soup.

There have been reports that George Washington underwent tooth transplantation procedures. But although tooth-transplantation advocate Jean-Pierre Le Mayeur served as Washington's dentist for a time, there is no evidence that he ever transplanted a human tooth into Washington's mouth. Dentures, though, were a different story.

I HAD BEGUN researching George Washington's dental woes by looking into the origin of the famous wooden-teeth story and had learned that the definitive source of the oft-repeated myth remains a mystery. There

---

* For a review, see Blackwell, "'Extraneous Bodies.'"

was no mention of wooden teeth in early biographies of Washington published in 1806 and 1855, but it was in wide circulation by the time of the George Washington bicentennial celebration in 1932.

In 1976, Washington-denture expert Reidar Sognnaes proposed the first modern-day hypothesis to explain how the tale might have originated. After noting the peculiar staining pattern seen on some of the teeth used in Washington's dentures, Sognnaes did his own experiments. A combination of "tobacco, tea, coffee, port wine, etc.," he wrote, coming into long-term contact with the teeth, led the public, who viewed the dental appliances a century later, to the belief that they were made of wood. The dark staining would have occurred primarily because the ivory or animal teeth used to construct his dentures had to be carved and filed down until they resembled human teeth. During the process of filing these teeth, it is likely that a portion of their mineral-hardened, protective covering of enamel would have been removed. The lack of enamel would have made the manufactured teeth more porous and so more likely to absorb substances like red wine. Sognnaes believed that the abundance of microscopic channels, called dentinal tubules, running through the teeth would have permitted entry of stain-producing substances (especially the port wine Washington favored) deep into the Founding Father's touched-up tusks.

Although Sognnaes's explanation for the misunderstanding about George Washington's wooden teeth seems reasonable enough, the story of his dentures would grow far more complicated beginning in 1999. Washington's second set of dentures was a full set, whose construction, like the older set, was also attributed to John Greenwood. The upper and lower portions were connected by two steel springs mounted at the rear of a pair of lead bases that had been cast to fit onto Washington's jaws. The tension produced by the springs kept the contraption in place (at least in theory), but it also required that Washington continuously contract his jaw muscles to keep his mouth from popping open.

A take-home message at this point, and one which will become more important later, is that wearing dentures in the eighteenth century was a painfully unpleasant chore that required constant attention and effort. I'm told it remains a hassle to this day, though perhaps not quite as severe.

Although Washington's spring-hinged contraption was far more intricate than the earlier partial models he had worn, the composition of the teeth affixed to it remained a mix of animal and human. Sognnaes determined that the upper set had been carved from the lower incisors of a five-year-old cow (and not an elk, as previously thought). The lower set contained a single tooth made of ivory. The remainder of the teeth, he identified as human in origin.

In the late 1990s, while researching the enslaved people at Mount Vernon, research historian Mary V. Thompson found a peculiar notation made in May 1784, in a ledger book from George Washington's Mount Vernon plantation—parts of which appeared in an article she wrote in 1999. The original entry had been made by Washington's distant cousin Lund Washington, who happened to be the manager of the Mount Vernon plantation at the time. Under the heading, "Cash p[ai]d on Acc[oun]t of Genrl. Washington," the notation read: "To p[ai]d Negroes for 9 Teeth, on acc[oun]t of the French Dentis [sic] Doctr Lemay [sic]."

The transaction was also transcribed into Washington's ledger of accounts as "By Cash pd Negroes for 9 teeth on Acc[oun]t of Dr. Lemoin."

"Dr. Lemay" and "Dr. Lemoin" were references to Jean-Pierre Le Mayeur, the French dentist who had a long and friendly relationship with George Washington and his family. The amount paid for the nine teeth was 122 shillings, or about 6 pounds, in colonial currency.

The information garnered little attention for nearly two decades, until . . .

GEORGE WASHINGTON'S TEETH NOT FROM WOOD BUT SLAVES
—*Philadelphia Tribune* (February 24, 2018)

YES, GEORGE WASHINGTON USED TEETH FROM ENSLAVED AFRICANS
—*Atlanta Black Star* (February 27, 2018)

GEORGE WASHINGTON DESPERATELY TURNED TO DENTURES MADE OF HIPPO IVORY AND 'SLAVE TEETH' IN A BID TO CURE HIS AGONIZING DENTAL PAIN
—*Daily Mail* (July 5, 2019)

STORE YANKS SOUVENIR WASHINGTON DENTURES OVER SLAVERY LINK
—AP News (February 19, 2020)

The upshot from these articles and the online barrage that followed was that George Washington had purchased teeth from enslaved people, and then had his dentist, Le Mayeur, use them to construct dentures. As a result, it is now generally accepted by the public that George Washington's teeth *weren't* made of wood but likely came from the mouths of people he held in bondage.

It's important to note that beyond the absence of modern dental care, the widespread increase in sugar consumption had created an even greater market in which the poor and enslaved, including those held for decades by George Washington, commonly sold their own teeth. Until the mid-1660s, only the rich could afford sugar, most of which came to Europe from India, and was imported and exported through Venice, Italy. Instead, those seeking to satisfy their sweet tooth did so with honey, fruit, and dates. Once sugarcane plantations, tended by hundreds of thousands of enslaved Africans became established in the West Indies, the price of "white gold" (as sugar was known to British colonists) dropped and its availability increased. Although oral bacteria, or any bacteria for that matter, hadn't been discovered yet, the microbes living in the warm, moist, and hospitable environments offered by the human mouth must have loved the news.*

Since few Westerners back then actually brushed their teeth, the bacteria thrived unchecked, behind thick, hardened walls of plaque, which built up to shield them. This cement-like substance was composed of salivary proteins, calcium phosphate, and the cellular secretions of the bacteria themselves. Except in instances when the plaque might have been removed by a dental instrument known as a scaler, the bacterial horde would remain effectively walled off from outside interference.

Interspecies squabbling was not an issue, since the microbes had already divvied up the mouth into environmentally varied niches—some inhabiting the so-called supragingival plaque, located above the gumline, others camped out in the dark and foreboding subgingival caverns, situated below the gumline and tooth crown (the exposed portion of a tooth), still others teeming on the surface of the tongue. There, they were "born," grew, reproduced, and died. During breaks in the action, they reproduced some more.

The fuel for all that activity came from the chemical breakdown of the incoming sugar, which released ATP, the microbes required to keep things exciting. Unfortunately for the unsuspecting bipedal hosts, the bacterial feeding frenzy taking place in their mouths released tooth-dissolving lactic acid as a byproduct.

Contributing to the sugar-stoked dental mayhem was a concurrent

* Or whatever passes for love in the single-celled set.

increase in the consumption of carbohydrate-rich milled grains, which (unlike today, thankfully) included a good measure of stony grit, shed from the millstones used to grind the grain. This microgravel was baked into bread and other consumables. During the chewing process, the grit further wore down the tooth's superhard and protective enamel layer, eventually exposing the bone-like dentin* below it to rapid and devastating wear and tear.

The breakdown of the enamel also provided further inroads for the bacteria. The result was a seventeenth- and eighteenth-century explosion in the number of dental caries, which is when a new breed of professionals sprang up to deal with the upswing in tooth-related issues. The bacteria were reportedly less than pleased.

Although George Washington and his teeth began parting ways at an early age, sugar may not have been the primary reason. Colonial America historians and Washington researchers alike have suggested that the reasons behind his long-term dental problems included poor diet, primitive dental care, gum disease, loss of bone associated with the tooth sockets (i.e., alveolar bone resorption), and genetics. There is also a good chance that Washington's prior medical history included treatments with what would now be regarded as harmful substances, including some that were injurious to teeth. Key among these toxic treatments was the substance known as *argentum vivum* or quicksilver.

Better known today as mercury, the silvery-looking element is the only metal that exists in liquid form at room temperature. Mercury is now infamous for the health risks it presents when consumed, absorbed through the skin, or inhaled as a vapor, but it was formerly widely employed, from the sixteenth century to the twentieth century, as a

* It's spelled "dentine" in the United Kingdom.

therapeutic, commonly used to treat syphilis, parasites, and influenza. In its powdered form, calomel (mercurous chloride) was thought to heal malaria, yellow fever, smallpox . . . and teething pain in infants.

Whether medicinal or environmental in nature, mercury poisoning has been implicated in a slew of health issues. Many of them, like tremors, depression, and memory problems, are neurological.* But they also include the dental symptoms experienced and vividly described by one of Washington's fellow Founding Fathers. According to John Adams, who had undergone a regimen of "Milk and Mercury" to treat a bout of smallpox, "every tooth in my head became so loose that I believe I could have pulled them all with my Thumb and finger." It was, the second US president claimed, a condition that "rendered me incapable . . . of speaking or eating in my old Age, in short they brought me into the same Situation with my Friend Washington."

Washington had also been treated for smallpox, having contracted the often disfiguring and frequently deadly viral disease in 1751. And although there is no record of Washington's smallpox-related medical care at the time, mercury-induced purging of the lower digestive tract with calomel was the typical treatment method. It was also a treatment that Washington knew well. He would suffer through its side effects for much of his adult life—all in the mistaken effort to balance the four humors.

Thought to be the all-important quartet of bodily fluids responsible for health and sickness, blood, black bile (which does not exist), yellow bile, and phlegm (the smoker's favorite) had been dominating the field of medicine since the Roman physician Galen (129–216 CE) wrote

* The term "Mad as a hatter" did not originate with Lewis Carroll's famous character but from hat factory workers in the eighteenth and nineteenth century exposed to the mercury they used to soften the material used to produce hats. Symptoms of long-term exposure to mercury vapors included tremors, psychosis, hallucinations, and irrational speech and behavior.

about them in the second century, after cribbing them from the ancient Greeks. Galen's voluminous and often wildly mistaken assertions about the human body would effectively hamstring Western medicine for over fifteen hundred years. As a result, therapeutic bleeding and purging became the "Take two aspirin and call me in the morning" of its day.*

Note: If the relationship between teeth and mercury rings a bell, it's likely because of the controversy regarding amalgam—an alloy of mercury that was used to fill the space formerly occupied by drilled-out dental caries for over 150 years. Arguments about the pros and cons of mercury-based amalgams would escalate into what would become known as the Amalgam Wars of the mid-nineteenth century, whose primary victim would turn out to be the American Society of Dental Surgeons, the first organization attempting to standardize dentistry in the United States. But more on *that* conflict later.

Getting back to the question of whether or not Le Mayeur had constructed George Washington's dentures using the teeth of enslaved people—I called Susan Schoelwer, senior curator at Mount Vernon, about the entry in Washington's ledger from May 1784. The record of the sale of these nine teeth had become the key piece of evidence that these very same teeth had somehow found their way into Washington's mouth.

Schoelwer began with what seemed to me to be a bit of superfluous information—albeit information that I had not heard before. She said that in keeping with bookkeeping practices of the day, the transaction in question had been recorded in two different places: the "journal of accounts," a simple list of payments and receipts as they occurred, and

---

* Admittedly, since purging resulted in the explosive emptying of the lower digestive tract, the purging/aspirin comparison isn't perfect. Mouth-related side effects of calomel treatment included irritated gums, mouth sores and tooth loss. Derived from the Greek for *kalos* (meaning "beautiful") and *melas* ("dark" or "black"), calomel *may* have been named for the dark stool it produced after ingestion. This was mistakenly thought to indicate the presence of black bile.

later in the "ledger of accounts," which was a more formal record used to organize information on a client-by-client basis. Schoelwer emphasized that the transaction was clearly recorded as being "on the account" of dentist Jean-Pierre Le Mayeur.

"What does that mean?" I asked.

"In the cash-poor economy of the American colonies, it was common for financial transactions to involve multiple parties: if A owed B, and B owed C, A could clear his debt with B by paying B's debt to C," Schoelwer explained. "In this case, the explicit notation 'on the account' of the dentist points to Le Mayeur as the end recipient of the nine teeth. If Washington had been purchasing the teeth for himself, there would have been no need for this information; the entries would have simply recorded the item and payment, as when Washington purchased poultry, wild game, fish, and garden produce from enslaved individuals."

As for why this may have been important, Schoelwer reminded me that at the time of the transaction, the French dentist was practicing in the Mount Vernon area and advertising in newspapers in nearby Alexandria, Virginia, his availability to perform tooth transplants. Notably, Schoelwer said, references to this procedure recommended "that dentists obtain a number of specimens, in order to maximize chances of matching the recipients' natural teeth," and because of this, "dentists who made dentures frequently maintained a stock of teeth as part of their professional tool kits."

According to historian Mary V. Thompson, "Slaves of the eighteenth century sometimes turned to the perfectly acceptable means of making money by selling their teeth to dentists."

Apparently, it was also "perfectly acceptable" to rip off enslaved people, paying them far less than the going rate normally paid for the teeth of the nonenslaved. A case in point was George Washington himself, who laid out less than one-third of what Le Mayeur had been offering (three guineas) for good front teeth "from anyone but slaves."*

So . . . although there is a possibility that Le Mayeur transplanted some or all nine of these teeth from enslaved people into Washington's jaws or employed them in a denture for the Founding Father, there is nothing remotely resembling proof that Le Mayeur did either. The sole evidence that does exist relies on speculation about the meaning of the Mount Vernon journal/ledger notations.

Given Schoelwer's "A owed B, and B owed C" explanation, we might hypothesize that the notations recorded a deal in which the teeth were *not* destined for Washington but were instead used to pay a debt to Le Mayeur. In other words, Washington (A), who owed Le Mayeur (B) for dental services for him and his family, grossly underpaid the enslaved people on his plantation (C) for nine teeth, then gave the teeth to Le Mayeur to pay off the debt Le Mayeur would have had with the slaves had *he* purchased the teeth himself.

It should also be emphasized that the two surviving dentures belonging to Washington that did contain human teeth were almost certainly constructed by a different dentist (likely John Greenwood) five years after the journal and ledger notations concerning Le Mayeur.† There is no indication of where these human teeth came from. Nor is there, of course, any indication that George Washington would have had any problem whatsoever had the teeth in those dentures come from enslaved people.

In the end, the only definitive evidence relating to this story is that

---

* Le Mayeur's "payment of three guineas for good teeth from anyone but slaves" is cited in "September [1785]," *Founders Online*, National Archives, https://founders.archives.gov/documents/Washington/01-04-02-0002-0009.

† It has been estimated that Washington consulted at least eight dentists over his lifetime.

(1) George Washington had considerable dental problems over his entire adult life; (2) he was, throughout his life, under the care of several dentists; (3) an unknown number of these dentists and associated dental artisans were involved with fashioning dentures for him (which included some human teeth), and these required frequent repair; (4) Washington held human beings as slaves (approximately three hundred of them at the time of his death in 1799), and on at least one occasion in 1784, he paid an unspecified number of them what was apparently a *miserly* amount to have a total of nine teeth pulled and transferred to French dental surgeon Jean-Pierre Le Mayeur; (5) Le Mayeur, who provided unspecified dental care to Washington in the 1780s, specialized in tooth transplants—a procedure Washington declined on the record, later expressing surprise when he learned of a successful operation.

As we've seen, the dentures fashioned during George Washington's time, especially full sets, were heavy (with bases composed of lead), difficult to control, and painfully uncomfortable to wear. It was a problem that led presidential artist Gilbert Stuart to later comment on his iconic Lansdowne portrait of Washington in 1796, "When I painted him, he had just had a set of false teeth inserted, which accounts for the constrained expression so noticeable about the mouth and lower part of the face." There are additional denture-related references written by Washington, and those who knew him, each of them pointing to a single conclusion: eighteenth-century dentures sucked.

Initially, I believed that the primary reason the rich and edentulous would suffer through wearing these ill-fitting and painful contraptions would be to dine in something resembling a normal fashion. I was surprised to learn, then, that dentists of the day recommended that dentures be removed before meals. The main reason for this was that the spring-hinged nature of these appliances made it impossible for the dentured to move their jaws from side to side—a masticatory necessity whose evolution turned the mammalian mouth from a snap trap (think alligators) into a far more efficient grinding mill (moo!). In addition, the tendency for these eighteenth-century devices to pop out at inappropriate times turned the simple act of chewing into an unpleasant chore, during which there was a constant threat of embarrassment and mechanical failure. By all accounts, though, Washington did wear his dentures while dining, as well as during the inevitable toasts that accompanied those meals—though he was reported to have eaten little or nothing during these get-togethers.

The reason Washington wore dentures that were uncomfortable and useless for eating had everything to do with his distinguished place in the newly minted American experiment in democracy. His ability to speak well and present a strong, attractive physical appearance were direct reflections of the nation's character—this, at a time when the new country could not afford to present itself as anything less than morally upstanding, smart-looking, and powerful. As commander of the Continental Army, and then as the first president of the United States, Washington was often called upon to address the men he commanded, crowds, and important leaders. This is something that would have been unthinkable without dentures, given his completely or nearly completely toothless condition. But even when sporting a full set of dentures, Washington's speeches were generally short in nature, which may reflect the difficulties of wearing them.

George Washington needed to look good, and dentures corrected what was at the time considered to be a disfiguring facial condition. According to historian Jennifer Van Horn, this was vitally important, because his lack of teeth would have been equated to a lack of moral character and the absence of self-control—both important requirements for a gentleman whose position required that he be held in the very highest esteem.*

UNFORTUNATELY, THE STIGMA associated with the perception of flawed appearance continues today. The American public has become far more demanding in its requirement for dental perfection. The quest for the Great American Smile likely had its origin in Hollywood during the late 1920s. Oral hygiene was a mess back then, and things would not improve until fluoride-containing toothpaste was developed in the 1950s. For example, from November 1940 through September 1941, 8.8 percent of young American military enlistees could not qualify for general service because of gum disease or for having less than twelve opposable teeth—and that included dentures.†

With the birth of the talkies in the late 1920s, actors and actresses were required to open their mouths, spout dialogue, and break into song. The problem was that they had to do so in front of movie cameras that would capture every detail of dental imperfection. Since the condition of a star's teeth could either make or break a career, studios sought the help of professionals—none more famous than dentist Charles Pincus (1904–1986).

Pincus used a mixture of porcelain and plastic to fabricate tooth caps (known as veneers) that could be fitted in place over existing teeth. These he used to fill gaps or cover up the often far-from-perfect teeth belonging to a roll call of Hollywood types that would make any autograph collector swoon. This list would come to include Jack Benny, Fanny Brice, Montgomery Clift, Joan Crawford, James Dean (who supposedly lost two

---

* A more detailed examination of the reasons why Washington used dentures can be found in Jennifer Van Horn's article "George Washington's Dentures."

† By October 1942, dental-related requirements for induction were nearly eliminated, and only 1 percent of those registering for military service were disqualified for gum problems and dental defects.

of his front teeth in a trapeze accident), Walt Disney, Bob Hope, Robert Taylor, and Mae West.

The Dentist to the Stars also fashioned a set of transitional teeth for Shirley Temple, thus saving the studio thousands of dollars in potential filming delays once America's Sweetheart began losing her baby teeth. According to medical journalist Mary Otto, though, Pincus's most lasting accomplishment was to provide veneers to a beautiful though gap-toothed young actress in 1939—thus helping Judy Garland define the bright-white American smile during her Technicolor adventures over the rainbow.

A FINAL NOTE related to George Washington and teeth concerns the involuntary nature of the toil and hardship experienced by millions of enslaved people in the United States. It stands in stark contrast to the benevolent portrayal usually afforded the first president, since, like other enslavers, Washington used descriptions of their teeth in advertisements seeking to recover enslaved people who tried to escape his Mount Vernon plantation. For example, from August 11, 1761: "*Neptune*, aged 25 or 30 . . . thin jaw'd, his Teeth stragling and fil'd sharp" and "*Cupid*, 23 or 25 Years old . . . round and full faced, with broad Teeth."

There is strong physical evidence as well that dental disease was widespread among enslaved populations in the antebellum South. We know this primarily from the growing field of paleopathology, in which researchers learn about diseases in past populations by studying preserved hard parts like teeth and bones. In the case of enslaved people, the lack of dental care and effective dental hygiene was one factor. Another factor contributing to dental problems was corn consumption, especially typical in the diets of those working on plantations like Washington's, which emphasized the crop. With its unintentional additive of stony grit, it's no surprise that tooth loss was commonplace among enslaved adults.

Enamel is normally laid down in incremental growth lines that appear on tooth surfaces as tiny transverse grooves called perikymata. These perikymata undergo measurable changes due to disease, chemical erosion, and mechanical abrasion. They remind me of tree rings, which are formed when a new layer of vascular tissue grows between the trunk and the bark during each growing season.

Larger, thinner cells are laid down earlier in the season, when more water is available to be transported from the roots upward through the trunk, resulting in springwood, which looks lighter in color. Later in the summer, as the temperature rises and there's less water to transport, smaller and darker cells (which are denser) are laid down, leaving a concentric pattern of light and dark wood to the outside of the previous year's growth. These growth rings famously provide dendrochronologists with information about a tree's age and the conditions present as a tree grows.

I wondered if similar principles regulated the deposition of growth lines in teeth, and so I contacted Griffith University paleoanthropologist Tanya Smith. She explained that hard tissues like shells, scales, and plant parts, form through rhythmic secretions driven by physiological changes inside organisms, like differences in ambient temperature and water availability in plants. "In the case of teeth," Smith told me, "the enamel, the dentin, and the cementum all show secretory rhythms ranging from eight hours to a year. The shorter (subdaily, daily, and near-weekly) rhythms in enamel and dentin begin prior to birth and continue throughout childhood, and longer seasonal and annual rhythms in cementum form once teeth emerge into the mouth and throughout adulthood."

Growth lines can be used to determine whether malnutrition or health-related disruptions occurred during the growth of the tooth. Evidence of disease or a developmental abnormality is commonly indicated by the presence of regions called hypoplasias (Greek: "under" +

"formation"), where enamel is either missing or deficient, leaving pits or ringlike indentations that wind around the circumference of the tooth. Hypoplasias have been recorded in teeth from individuals dating back through the preindustrial period and into prehistoric times.

Paleopathologists examining the disinterred teeth and bones of enslaved Americans have also learned that a large percentage of them died young—*very* young. Ohio State University historian Richard Steckel has concluded that the health of newborns and young children within enslaved populations "was comparable to that in the poorest populations ever studied" and that they died at rates of 350 and 200 per thousand, respectively—or about twice the rate determined for the same age groups for nonenslaved people across the country.

Those who did survive childhood often suffered from infectious diseases like tuberculosis and congenital syphilis. The latter, passed from mother to child in utero, can be diagnosed centuries later by the presence of so-called mulberry molars (with multiple rounded rudimentary cusps) and Hutchinson's teeth (barrel-shaped, notched incisors).

Additional dental defects have been used to help diagnose other congenital diseases within enslaved populations. For example, sickle cell disease (SCD) is caused by a mutation in the gene responsible for producing the oxygen-carrying pigment hemoglobin (Hb). This mutated form of hemoglobin (HbS) causes a physical distortion of the erythrocytes (red blood cells) that contain it, twisting them into the characteristic sickle shape that gives the disease its name.* Not only are these sickle cells inefficient at carrying oxygen (which is pretty much the sole function of red blood cells), but they tend to stick together and clump. In doing so, these

* Normal red blood cells are described as biconcave disks. To picture this, imagine a hollow rubber ball (we used to call them stickballs) and place it between your thumb and index finger, then squeeze those fingers together tightly. What you're holding is a rough approximation of the shape of a red blood cell.

cells form clots that block the entry of blood into smaller vessels. This cuts off the blood supply to the tissues and organs located downstream from the blockage and leads to a long list of problems, including pain, chronic organ damage, and eventually death. Among SCD-related dental problems is that the sticky, mutant erythrocytes impair blood flow into and out of the teeth (via the previously mentioned apical foramen at the root tips). Ultimately, this leads to the death of the tooth pulp and eventually the entire tooth.

Although evidence of most symptoms of SCD disappear as soft tissues break down after death, clues remain behind in the bones and teeth. As with hypoplasias caused by TB or syphilis, researchers can detect a characteristic deficiency in the minerals that make up dentin and enamel. Furthermore, their diagnosis of SCD can be strengthened through the identification of teeth with atypically shaped roots. This condition, known as hypercementosis, results from the abnormal deposition of excessive cementum on the tooth roots.

TEETH ARE FREQUENTLY the only remaining physical evidence of their former owner's existence, and so they are of paramount importance, enabling researchers to piece together the often-intricate puzzles presented to them by the distant past. But before heading down that road, let's continue with some additional aspects of a unique form of torture, otherwise known as ancient dentistry.

[15]

Jaw Jewelry, Pliers, and Pelicans

Some tortures are physical and some are mental,
But the one that is both is dental.

—Ogden Nash

Is it safe?

—Dr. Christian Szell in *Marathon Man*

For most of the first millennium BCE, the Etruscan civilization stretched from just beyond the Po River in northern Italy, south through Tuscany and the western coast, and into the current Campania region on the shin of the Italian boot. Unlike the Italic tribes, who would eventually wrest power from the Etruscans beginning in 510 BCE, the Etruscans did not speak Latin. As such, they had a range of customs that were very different from those of the ancient Greeks and the upstart Romans, whose own empire sprouted from an enclave of small villages and blossomed into a mighty republic.

Of course, the Roman Empire grew in famously blood-drenched fashion, and as would become something of an ugly tradition with these sorts of things, as the new guys displaced the old guys, they then trashed many of the old guys' customs. It was a task likely made easier by the language differences. Primarily because of this, there are no known surviving texts in the Etruscan language. One can imagine Roman men, though, directing rude hand gestures at the degree of authority, freedom, and social integration previously experienced by Etruscans of the opposite sex. For example, Etruscan women apparently ate meals with the menfolk, which was a big no-no for the Greeks and Romans.

What the Romans did crib from the Etruscans were many elements of Etruscan technology—especially their skill as metallurgists. The archaeological evidence is relatively rare, though, much of it uncovered below the ruins of the ancient Roman structures that had been built over them. But what it does clearly show is that the Etruscans had a thriving civilization, with great cities, beautiful art, and a complex social structure.

I spoke to Marshall Becker, emeritus professor of anthropology at West Chester University. An expert on ancient dentistry, Becker coauthored *The Etruscans and the History of Dentistry.* I was interested in just those topics.

By way of a preface, Becker explained that he and his colleagues have faced problems while trying to clarify the history of dental appliances and the replacement of teeth. These had to do with authors who wrote about the relics without seeing them, copies of these devices being treated as originals, mistaken information about supposed Egyptian and Babylonian dental contraptions, and poor recordkeeping regarding the relics that had been discovered. While there are glyphs and papyri detailing medicines and medical procedures, there is no written evidence, to suggest that dentistry played an important role in either Roman or Babylonian civilizations (though there is some physical evidence that ancient Romans did extract teeth). Likewise, for the existence of Egyptian or Babylonian false teeth and dental bridges in anybody's moldering mummy mouth.

It was Etruscan goldsmiths, Becker said, who constructed the very first dental appliances, and they did so for a very specific clientele and for reasons that had nothing to do with medicine or dentistry. He explained that he and his colleagues were ultimately able to separate the truth from the misinformation because of a small set of original devices whose characteristics told an interesting story. Intrigued, I asked him to fill me in.

According to Becker, these appliances, the earliest dated to 600–700 BCE, were distinctly Etruscan in design, rare, and created solely for wealthy women. But instead of portraying strength and vitality, or correcting a mouthful of dental health issues, Etruscan dental appliances were worn solely to advertise the wealth and lofty societal status of their owners.*

The process began with the removal of all four upper incisors, probably without anesthesia. Artisans then filed down the roots of these teeth and reshaped them for use in the appliance. Unlike the bulky eighteenth-century dentures, where modified human and animal teeth were wired to a heavy lead base, each of the modified Etruscan incisors was secured by tiny rivets between a narrow band of gold that had been folded in two. The ends of these bands were fashioned so that they could loop around the adjacent and unextracted upper canine teeth. Since the canines remained rooted to the upper jaw, these teeth secured the gold band in place (see figure 34).

I asked Becker how he'd determined that the gold bands were not designed as functional replacements but as ornaments.

He told me that studies on the size of the teeth involved indicated that they had come from young to middle-aged women, and that upper incisors were not the types of teeth that were typically either lost at an early age or wore out quickly in adulthood. Several studies, including one by Becker and his colleagues, have demonstrated that Etruscans had "rather good dental health, and a pattern of tooth loss that begins only in

---

* False teeth could have had an even earlier origin. The oldest known dental prosthesis may date to 7000 BCE. It consisted of a carved bone that had been inserted into the space where an upper molar had previously been.

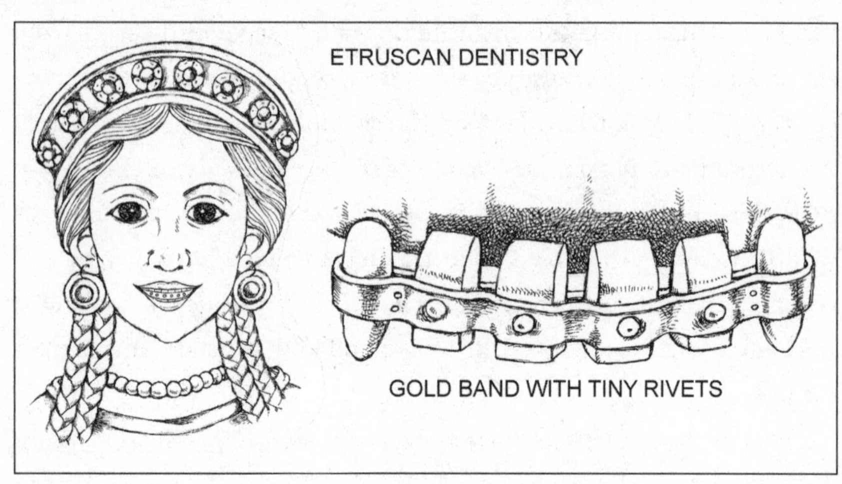

FIGURE 34

late middle age," when "dental loss invariably begins in the molar area, and only gradually affects the anterior dentition." This led them to conclude that the appliance owners had undergone "tooth evulsion"; in other words, they'd had their healthy teeth yanked—in this case to have them refitted as filed-down versions bound in gold.

The fact that the tooth roots had been eliminated provided additional evidence that the contraptions would be useless (or close to useless) for chewing food. Basically, then, they were constructed and worn for show.

After I posted a photo of an Etruscan dental appliance online, one of my friends, the neurophysiologist and science communicator Kiki Sanford, compared them to the "grills" worn by the entertainer Madonna starting around 2013. A bit of investigation left me relieved to learn that the Material Girl had not had her upper tooth row pulled. Instead, her gold- and jewel-laden contraptions fit over her existing teeth and could be easily removed.

The Mayans did something similarly ornamental. With a probable origin sometime during the first few hundred years CE, their technique entailed drilling the lip side of the front teeth—likely with bow drills—and then filling the holes with inlays made of finely crafted stones like jadeite and pyrite (see figure 35).

It is believed that the Romans, Greeks, Egyptians, and others in the Near East adapted some Etruscan dental technology for use in nonornamental functions—using copper, bronze, or gold wire braces to stabilize defective or loose teeth (though there is little physical evidence). It was a practice that continued until dental appliances began to appear in medieval Europe.

Unlike the Etruscans, the Romans *were* involved in therapeutic tooth pulling, as evidenced by the eighty-six mostly cavity-ridden molars that had been yanked from the aching jaws of fifty or so individuals (primarily

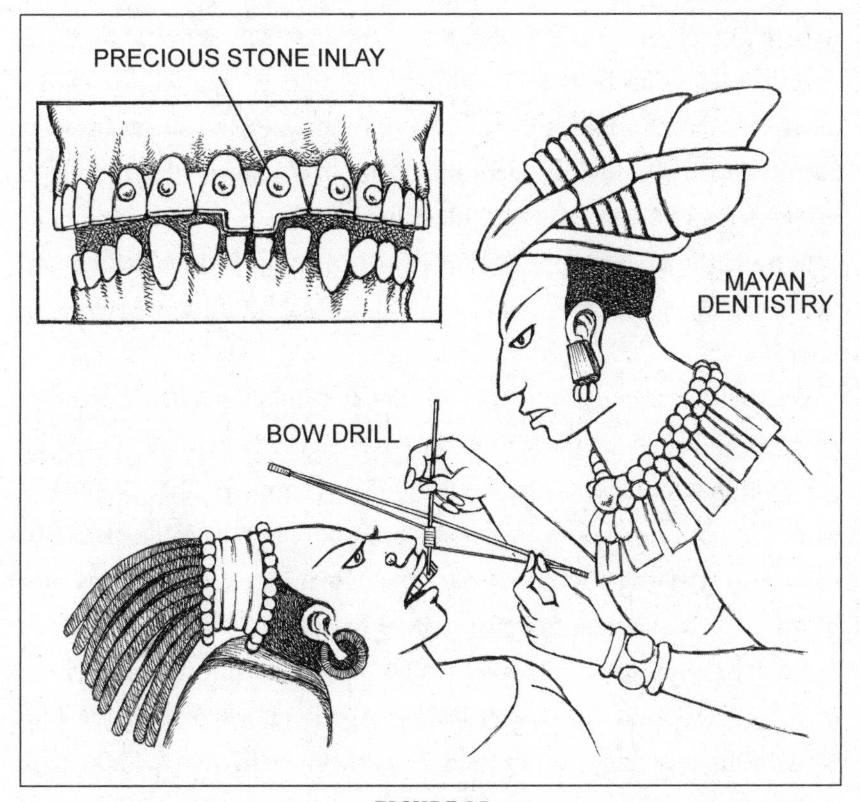

FIGURE 35

men, this time around). The teeth were found deposited in the drain of a *taberna* (shop stall) within a series of recesses built into the Temple of Castor and Pollux, which was uncovered in the ruins of the Roman Forum, and dated to somewhere between 50 and 110 CE.

At approximately the same time, Emperor Claudius's physician, Scribonius Largus, and others described dental procedures like the use of forceps and the cutting down of gums to ensure that the roots of the teeth they extracted were not left behind, something they knew could lead to serious problems.

As for the ancient extractions themselves, a number of tools not designed for that purpose—pliers, forceps, and the like—were co-opted for use in pulling teeth. But while there is evidence that dental extractions were taking place as early as the second century BCE, it would be many hundreds of years before specialized instruments were crafted solely for use in these procedures, and it would take nearly as long for metalsmiths to begin making dentures. "The first known illustrations of instruments that can be clearly identified as used in dental extractions date from the medieval period (as in the images of Saint Apollonia)," Becker and his coauthor, archaeologist Jean MacIntosh Turfa, wrote in their book.

In 2022, I saw one of these, a painting by Guido Reni, dated to 1600–1603 CE, on loan from the Museo Nacional del Prado in Madrid and displayed at the Borghese Gallery in Rome. I was stunned to see that Reni's work had nothing to do with a dental procedure but instead depicts two men, each wielding a tool that resembles a blacksmith's pincers as they torture Apollonia, who has been bound to a post. The young woman's eyes roll heavenward as one creep brings his pincers threateningly closer to her mouth. Somewhat less obvious is Apollonia's second abuser, who with one hand is yanking her head up by the hair (heedless of her halo). In his other hand he is gripping his own set of pincers, the business end of which holds a previously extracted tooth. Shocked, I couldn't resist pointing out the tooth to a fellow tourist, who shot me a strange look before hurrying away.

I would later learn that Apollonia's torture took place in 249 CE and ended with her being burned alive. As a result, Apollonia, who had run afoul of anti-Christian sentiment in Alexandria, Egypt, became the patron saint of dentists and those suffering from toothaches.

Among connoisseurs of early dental gear, perhaps the most famous instrument was a nasty-looking hooked lever known as a pelican (see figure 36), so named for its vague resemblance to the beak of the baggy-throated waterfowl. Medical historian and author Richard Barnett wrote that before becoming the go-to instrument for tooth pullers, pelicans were used by coopers to press metal hoops onto the wooden staves they used to build barrels.

The tool consisted of a main shaft, sometimes equipped with a handle at one end, and a metal arm or two riveted to the center of the shaft. During an extraction, the hooked tip of the arm (known as the claw) was placed on the tongue side of the offending tooth. At the end of the main shaft (as pictured) or, in some instances, on another arm, a semicircle of serrated metal called a bolster was wedged against the outside of the patient's jaw.* While immobilizing the patient's head, the dentist pushed downward on the pelican's handle, which would (in theory) lever the tooth out of its socket.

With an initial mention as a tooth-pulling device in 1363, the pelican of the nonavian variety went through several modifications between the sixteenth century and the eighteenth century to increase its efficiency and adaptability. These design tweaks included the addition of screw threads that permitted the dentist to adjust the instrument for different

* Instrument part names from "Dental Equipment," British Dental Museum, https://bda.org/museum/collections/dental-equipment.

teeth and variations in the size of a patient's mouth. But improvements or no, pelican-assisted extractions must have been gruesome and painful affairs in which the jaw, gums, and surrounding teeth might have been seriously damaged.

Even more dangerous during these times of nonsterile conditions were the tooth-related infections that reportedly killed more patients

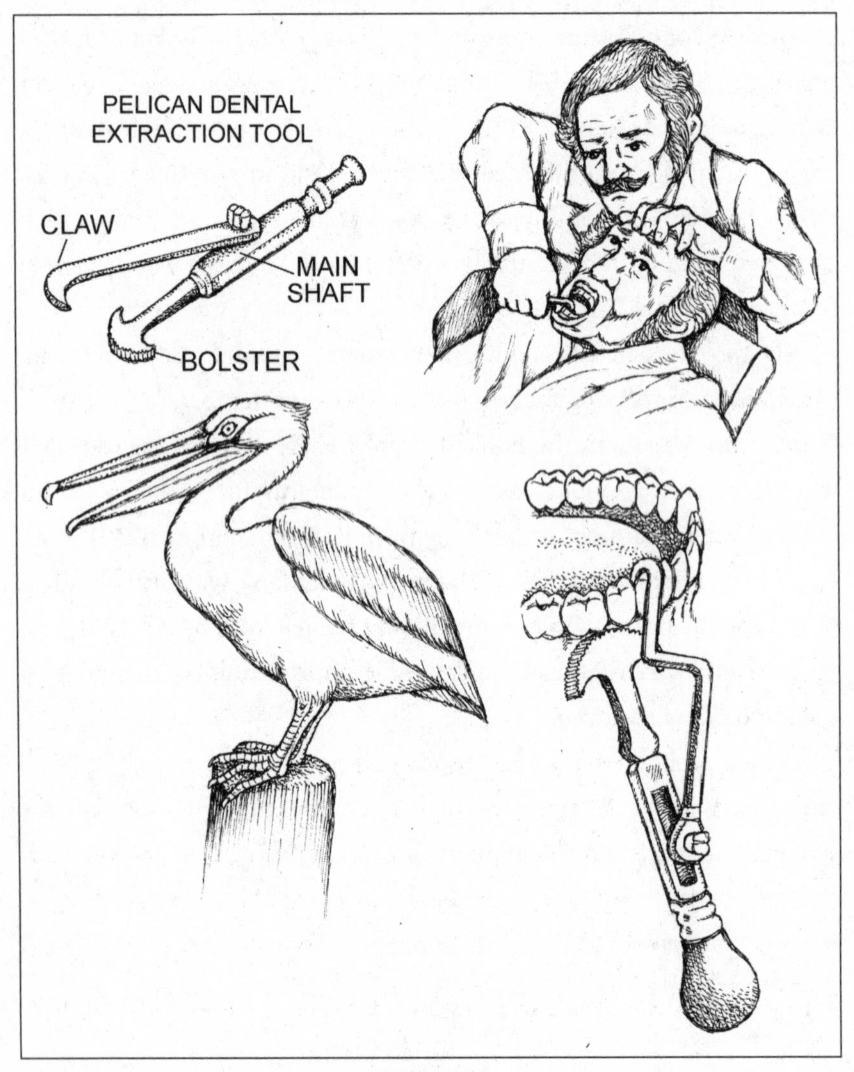

FIGURE 36

than seems possible—at least to me. Take for example the 111 people reported to have died in London from dental issues during a single week in the summer of 1665.

That number was so shockingly high that I asked my dentist, Kevin Sheren, how it could possibly be so. Always eager to talk about the anatomy of the mouths he's spent a career peering into, Sheren pulled up a set of my own x-rays and pointed to a molar on my upper tooth row. "Look how close those roots are to that sinus," he said, referring to one of the thin-walled hollows found within some skull bones. "On the other side of that sinus is your brain." An untreated infection in the tooth, he explained, could easily spread from tooth root to sinus and then on to the brain. "The next thing you know you've got a life-threatening brain infection."

I wondered if a problem with the lower tooth row could also be deadly, and the tooth meister said, yes, that bacterial infections like those from an abscessed lower molar, sometimes spread from the lower jaw to the floor of the mouth.* From there it's on to the pharynx and the neck, where the infection speeds, highway-style, along the connective tissue covering the carotid artery and jugular vein.

"Giving it a clear pathway into the chest cavity," I offered.

Sheren nodded and went on, perhaps a little too eagerly for some. The next stop on the Bacterial Express is often the mediastinum, a fleshy compartment running down the middle of the chest cavity. In addition to the heart, the mediastinum also contains the thymus gland and parts of the trachea and esophagus. It's here that swelling (edema) from

---

* A dental abscess is a pus-filled bubble within soft tissue like the gums or the tooth pulp, the innermost part of a tooth, which contains nerves and blood vessels. As for pus, think if it as the corpse-littered battlefield that forms during a confrontation between an infective agent (like disease-causing bacteria) and the body's defense team (led by several types of white blood cells). During an infection, as the bacteria-laden pus bubble swells, the toxins released by the bacteria and the chemicals released by cells of the beleaguered home team stimulate pain receptors. Once the abscess bursts, the pain may be relieved, but the infection remains and can spread further.

a fast-moving infection known as descending mediastinitis can lead to problems with breathing and swallowing, and even death.

Though it is impossible to determine just how many of the seventeenth-century Londoners populating the *Bill of Mortality* died from such infections or at the hands of charlatans and untrained dental practitioners, in all likelihood the numbers were considerable. As for the latter, a gradual understanding of the dangerous consequences of non-sterile conditions, combined with the development of more efficient dental surgery and aftercare, eventually led to a significant decrease in the occurrence of tooth-related fatalities.

As many of us are all too aware, another common dental practice is having a tooth filled, a procedure in which a cavity is repaired by first removing the damaged tissue (usually with a drill), then placing filler material into the resulting hole. The Chinese appear to have been the first to document the use of blended metal substances called amalgams for this purpose, with the earliest mention from a medical text by Su Gong (also known as Su Jing) written during the Tang dynasty in 659 CE.

The formula for an amalgam of mercury, tin, and silver was published in 1505, and its use in China was commonplace by the end of the sixteenth century. It showed up in Europe soon after. Recent studies support the notion that the European blend was the actual or a modified version of the Chinese recipe, though it's unclear just how amalgam arrived in Europe. Variations on its formula centered around mixing shavings from silver coins, copper, and tin with liquid mercury, forming a workable mass that could be inserted ("condensed") into the drilled-out tooth hollow.

Controversy over the safety of amalgams arose in the nineteenth century. Some dentists believed that the mercury-containing substance was safe, and others warned of the danger of mercury poisoning. As an alternative, antiamalgamists promoted the use of gold foil or other substances as a restorative material.

In 1840, several years before the arrival of mercury amalgam in the United States, the American Society of Dental Surgeons was formed to institute standards and requirements related to who could practice dentistry. Weeding out charlatans and quacks became a prime directive. In 1845, the first Amalgam War broke out, when the ASDS declared that all its members must sign a pledge denouncing the "use of amalgams for dental practices as malpractice." The uproar that followed eventually led to the disbanding of the ASDS in 1856 and the founding of the American Dental Association three years later.

Beginning in the 1960s, resin-based composites began to replace amalgam fillings. Initially, the composites consisted of two pastes that hardened epoxy-style when they were mixed. Some of us might recall our dentist mixing up this tooth-colored glop before ladling it into the hole they'd just drilled (and the pain we felt when the hole was first cleared with a blast of air). But given the short time that the dentist had to work on the mixture, this method gave way to curing (i.e., hardening) with light. Initially, ultraviolet light was used, but eventually wavelengths from the visible part of the spectrum were employed. This explains that blue-lit wand that dentists and their assistants often wave around in your mouth just after they've filled a cavity.

Though composites had pretty much replaced the use of amalgams by the early 1980s, amalgam skirmishes continue today, with supporters claiming that there are no significant health concerns for those (like me) living with a mouthful of old fillings. They argue that mercury in its chemically bound form is harmless and that there is no scientific evidence that these fillings cause any problems. Opponents implicate amalgams in health-related issues ranging from allergic reactions to mental disorders. These dentists advise their patients to replace old fillings with composites, which have admittedly gotten stronger, polish easily, and come in various shades of white and yellow to match the surrounding tooth material.

But as Centers for Disease Control and Prevention dental researcher Mary Ellen Mortensen wrote, "Whichever camp you follow . . . undoubtedly you can agree with the opposition that mercury serves no healthful purpose in humans. The smallest possible exposure is probably the most desirable."

AN EVEN MORE recent controversy has arisen over fluoride, a mineral that occurs naturally in rocks and soil, and unnaturally in some drinking water and most toothpastes. The story began in 1901, when dental school graduate Frederick McKay moved to Colorado Springs to open his first practice.* He immediately began seeing large numbers of locals with teeth that appeared to be permanently stained a splotchy chocolate brown. Slightly more than skeptical when residents blamed the condition on consuming too much pork, McKay followed up by querying area dentists, who promptly responded with a display of synchronous shoulder shrugs. Fortunately, things perked up when he convinced colorfully named dental researcher Greene V. Black to visit Colorado in 1909 to help him investigate the condition.

The pair made two significant discoveries, the first being that the mottled enamel (as Black called it) was due to an issue only seen in children, since adults who didn't already have the condition never developed it. Their second significant finding was that the brown-stained teeth were far less prone to developing cavities than teeth in those folks unaffected by the condition. Although Black passed away in 1915, McKay continued

* Much of this story was adapted from the NIH's "The Story of Fluoridation," https://www.nidcr.nih.gov/health-info/fluoride/the-story-of-fluoridation.

his research into the condition, which had by then been christened Colorado brown stain.

In 1923, McKay learned that residents living just across the Rockies in Oakley, Idaho, exhibited identical tooth staining. They told McKay that the condition had started showing up in their kids shortly after a new warm spring had been tapped as a communal source of drinking water. McKay advised the townsfolk to abandon that particular pipeline, and amazingly—at least when compared to the likely reaction today—they listened to the scientist! Several years after shifting to a new water source, McKay had his proof: there were no cases of stained teeth in Oakley's new crop of youngsters. To his dismay, however, he was unable to find anything specific in the water that might have been causing the problem.

In the early 1930s, H. V. Churchill, chief chemist at aluminum industry giant Alcoa, read McKay and Black's reports and decided to test the drinking water in Bauxite, Arkansas, another town where the brown, mottled teeth were prevalent. Using more sophisticated testing methods, the researchers at Alcoa learned that the water had an extremely high concentration of fluoride, a mineral commonly found where warm springs or hot springs are present. Contacting McKay, Churchill asked him to send water samples from Colorado Springs and other sites where tooth discoloration had been seen previously.

Testing confirmed that the permanent brown stains on the Colorado Springs residents' teeth were indeed caused by high levels of fluoride in the wells and stream water there. The term "enamel fluorosis" was coined, thankfully replacing "Colorado brown stain."

By the late 1930s, researchers at the National Institutes of Health had determined that fluoride levels in drinking water of up to 1 ppm (parts per million) did not cause fluorosis.

Several years after that, Trendley Dean at the NIH steered research efforts in a different direction, investigating the cavity-preventing properties that had been reported by McKay decades earlier and determining the minimum level of fluoride that could still lower the incidence of tooth decay. Dean proposed a field test in which fluoride would be added to drinking water, and in 1945 he got his wish: a fifteen-year-long pilot study in Grand Rapids, Michigan, that monitored cavity development in roughly thirty thousand schoolchildren after the city agreed to fluoridate their public drinking water. The study found a 60 percent reduction in cavities—results that ultimately led to widespread water fluoridation across the United States and the inclusion of fluoride as an additive in toothpastes and dental rinses, effectively, most supporters claimed, turning tooth decay into a preventable disease.

Recently, though, some have called for the removal of fluoride from drinking water, which would impact approximately two hundred million Americans. They refer to new studies whose results center on two claims: (1) water fluoridation does not significantly contribute to cavity prevention when compared to areas where the treatment is not used (like most of Europe) and (2) treatment of a medical condition (i.e., cavities) should fall to the medical community—not the local water company. Additionally, those opposing fluoridation often cite a Canadian study published in *JAMA Pediatrics* in which pregnant women who lived in communities supplied with fluoridated municipal water (and who reported drinking tap water) gave birth to children who had IQ scores reduced by three to five points when they were tested at three to four years of age.

As one would expect, there is intense debate on the topic of fluoridation. Currently, most dentists and experts at the World Health Organization believe that health concerns over fluoridated tap water are not justified. That said, the list of antifluoridation proponents is currently growing. Readers should stay alert for further developments—by monitoring the science-based studies on both sides of the argument, while weeding out the zealots and antiscience quacks. (Hint: as with similar science-related issues, my suggestion would be to look for a list of references at the end of an article and be skeptical of articles that don't have one, and keep that eyebrow similarly raised when scrutinizing blogs and social media posts.)

[16]

Tooth Worms

It would be a wearisome and unprofitable task to catalogue the almost numberless examples of a belief in the worm theory of dental disease in the works of medical writers and in popular folkloristic medicine.

—B. R. Townend, "The Story of the Tooth-Worm"

Perhaps the most intriguing tooth-related myth (with apologies to the tooth fairy) is the enduring and widespread belief that cavities, and the pain associated with them, result from the nefarious action of tiny creatures known as tooth worms. There is some degree of agreement among dental historians that the first depiction of these creatures dates to ancient Egypt. In 1944, one of these, medical and dental folklorist B. R. Townend, gave an approximate date of 1200–1100 BCE, which corresponds with more recent papers on the topic.

Belief in tooth worms lingered in some places into the twentieth century (and perhaps even the twenty-first), and I was just as surprised by their nearly worldwide popularity—though "infamy" might be a better descriptor. Having plagued the ancient Egyptians, tooth worms burrowed their way into the cultures of ancient Assyria and Rome, where they surfaced in the fifth and second centuries BCE, respectively. They became even more widespread during the Common Era, and in the Middle Ages references to tooth worms can be found in England, Germany, and Italy.

In the Middle East, the polymathic Persian physician Avicenna mentions them in his famous *Canon of Medicine*, completed in 1025.

Townend also found written and linguistic evidence of "the worm theory" of dental disease in India, ancient and mid-twentieth-century China, Madagascar, and Guatemala. According to Townend, nineteenth-century Cherokee rituals against toothache described a wormlike intruder wrapped around the teeth, and moving within and between them, while in the Philippines, some Indigenous people believed that consuming grub-filled wild plums would lead to unwelcome wigglings and twistings of the dental variety. This was similar to the idea proposed by Italian physician and medical writer Carlo Musitano in the late seventeenth century. He suggested that tooth worms arose "from the eggs of flies and other insects" laid on soon-to-be-consumed food.

In the eighteenth century, *The Tooth Worm as Hell's Demon* was created by an unknown artist in France. As illustrated in figure 37, it is a four-inch-tall ivory carving whose halves hinge open to show the torments of a toothache—depicted (in part) as a battle between naked club-wielding humans and a tooth worm—the struggle taking place amid a pile of skulls.

Townend turned to a hypothesis conceived by anatomist and Egyptologist Grafton Smith (1871–1937) to explain how a belief in tooth worms, as well as cultural practices like sun worship, the construction of megaliths (stone structures), deluge stories, embalming the dead, and circumcision, appeared in so many civilizations. Smith believed that these facets of what he termed "heliolithic culture" had but a single origin—in Egypt or Mesopotamia—and they were then spread by ancient mariners across pretty much the entire world, where they would have enduring effects on all other civilizations (including those in the Americas).

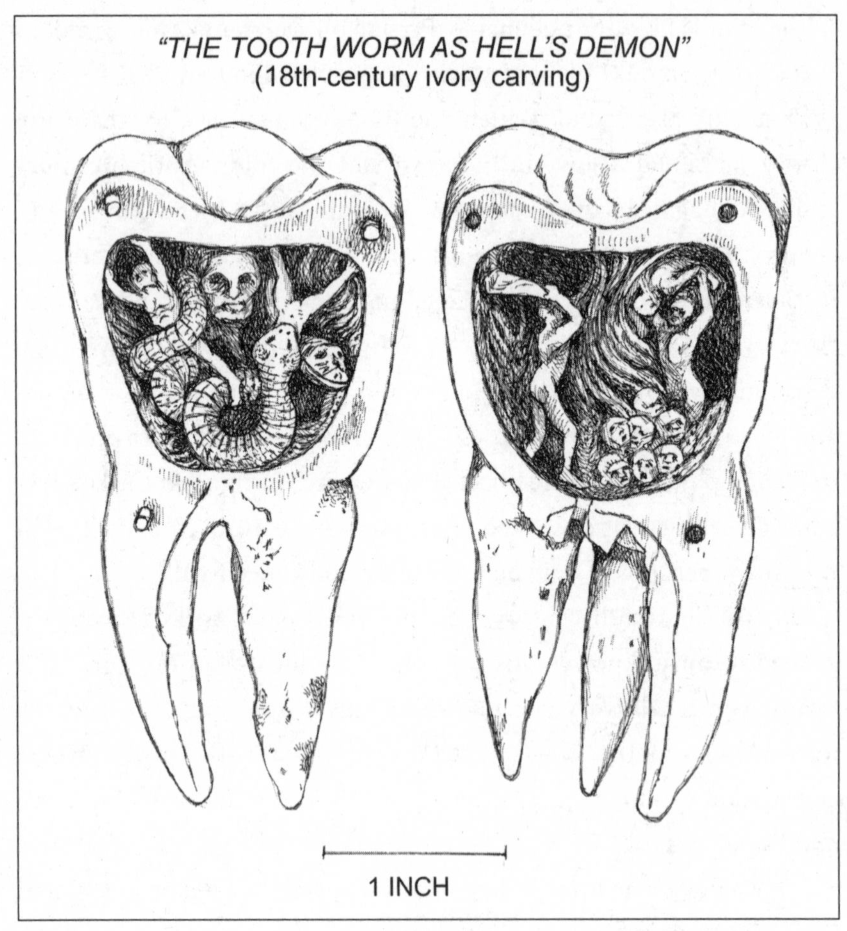

FIGURE 37

Today's archaeologists regard the concept of heliolithic culture and so-called hyperdiffusionism as simplistic and pseudoscientific—with no evidence that all societies derived from one predominant ancient culture.* But while more recent explanations dealing with expanding trade routes, migration, and colonization may explain the spread of some cultural phenomena, they do not fully explain the similarity of some ancient

* Well reviewed in Alice Beck Kehoe's *Controversies in Archaeology* (New York: Routledge, 2008).

beliefs between cultures that, initially at least, had no contact with each other. The New World references to tooth worms serve as a puzzling example, given the unlikelihood of dental-related cultural exchanges between the pre-Columbian inhabitants of Europe and South America.

Looking at this from an evolutionary perspective, in addition to the spread of practices between cultures, it seems equally probable that a belief in tooth worms may be the cultural equivalent of convergent evolution—that phenomenon in which similar traits or adaptations (such as wings) arise separately in extremely different animal groups (such as birds, bats, and pterosaurs), rather than diverging from a single common ancestor. As for the nonevolutionary version of convergence within modern humans, the phenomenon is better known among archaeologists and historians as independent invention.

It would have worked this way for tooth worms: Members of every cultural group, no matter where they were located and when they existed, would have seen firsthand the action of worms—whether it was the destructive boring of woodworms (the larvae of many beetle species) or the action of maggots (fly larvae) working within and upon decaying flesh. Additionally, insect-related diseases and worm-related parasitic problems, such as intestinal worms, plagued ancient humans (and they remain a problem in many places still). It is not a stretch, therefore, to suggest that far-flung and noninteracting cultural groups would look to worms as the cause of their tooth-related woes and then concoct treatments—some of them similar—to deal with them.

As for those cultures that *did* exchange information, treatments often followed those prescribed for other insect- or worm-related maladies. For example, since smoke was commonly used to combat swarms of flying pests like flies and mosquitoes, fumigation was also employed to smoke out tooth worms. The seeds of the flowering plant henbane (*Hyoscyamus niger*) were a popular fuel for this procedure as practiced by ancient

Romans, like Scribonius, who instructed toothache sufferers to place henbane seeds onto hot coals. Patients would then breathe deep what was determined centuries later to be the narcotic-laden (and quite possibly pain-killing) fumes before spitting out the imagined tooth worms.*

This smoke-assisted therapy was picked up by the second-century physician Galen, whose nearly three million words of error-ridden writings were still being followed more than fifteen hundred years after his death. The reason is that after the fall of the Roman Empire, Galen's voluminous work, like that of many Roman scribes, sat untranslated until the Middle Ages. At that point, Syrian Christian scholars translated it, but they did so into Arabic, giving Galen, who was a non-Christian monotheist, a decidedly Christian slant. Once the Arabic version was finally translated into Latin, church leaders throughout Europe loved it, going so far as to declare his writing to be divinely inspired.

Galen's inaccurate teachings spread throughout Western cultures, where physicians had to either follow by rote or suffer the consequences. Medical types quickly learned not to ask questions—and if they did, they asked them while church officials were off hunting witches. Given the potential for career setbacks like imprisonment, torture, and execution, most of those working in fields such as medicine and anatomy followed the doctrine of Galen until the seventeenth century—and in some places, far beyond that.

Historians Muwaffak Al Hamdani and Marian Wenzel wrote that in Iraq, the use of fumigation-based therapies to combat tooth worms continued until well into the mid-twentieth century. They suggested that the use of smoke to effectively combat swarms of mosquitoes in and around the city of Basra in the 1940s may have been adapted as a treatment for tooth worms. One such prescription had the toothache sufferer hold in

---

* Henbane (literally, "hen killer") is known to have toxic effects if consumed in large amounts.

a mouthful of cigarette smoke, which would supposedly kill the critters. In a related treatment, the afflicted would squat in front of a fire fueled by reeds and seeds while holding a dish full of water. After inhaling the smoke, tooth worms would reportedly fall from the achy mouth and into the water.

Over three thousand years ago, the ancient Chinese had their own beliefs regarding the existence of tooth worms, and their medical texts prescribed their own remedies to deal with them. ("Roast a piece of garlic and crush it between the teeth, mix with chopped horseradish seeds or saltpetre,* make into a paste with human milk; form pills and introduce one into the nostril on the opposite side to where the pain is felt.")

Therapeutic bleeding and cupping were also used to treat worm-related tooth pain. During the latter procedure, which is still practiced today, cups containing heated air are placed on the skin. As the air cools, it forms a vacuum, drawing blood and supposedly the bad stuff to the surface.

Another widespread remedy offers a perfect example of the Latin phrase "similia similibus curentur" ("like is cured by like"). Once again, it was tooth-worm maven B. R. Townend who dug up references to this approach. The treatments he unearthed included crushing caterpillars against a stricken tooth (ancient Assyria), gulping earthworms cooked in oil (ancient Greece), rubbing worms into a tooth cavity (Pliny the Elder, in ancient Rome), and fashioning an amulet containing worms and bread, then binding it to the sufferer's arm (ancient Greece again).

It is not a stretch, therefore, to see how these forms of annelid-assisted aid, and others like them, may have paved the way for the multibillion-dollar alternative-medicine industry known as homeopathy, in which being exposed to tiny amounts of whatever ails you is

* Spelled "saltpeter" in the United States and also known as potassium nitrate, this chemical compound is a key component of both fertilizer and gunpowder.

supposed to have curative effects. And if that sounds reasonable and perhaps similar to the science behind vaccinations, it really isn't—even though some of these alternative treatments are referred to as "homeopathic vaccines."

Consider that the most dilute doses of homeopathic medicine are said by proponents to work best. In fact, the most sought-after (and most expensive) dilutions are those that no longer contain *any* molecules of the original substance—just something supposedly akin to its chemical memory. This is a concept that has not been supported by a single peer-reviewed scientific paper. Vaccinations, on the other hand, work by stimulating the body's immune system to produce protective protein bits called antibodies. And unlike homeopathic treatments, vaccinations aren't swallowed, to be broken down by the digestive system. Nor do these alternative remedies possess antibacterial or antiviral protection, or produce immune system-generated resistance as seen in vaccines.

But whether you subscribe to the concept of homeopathy or not, the popularity of these "hair of the dog that bit you" types of tooth-worm remedies continued into recent times. Medical historian Werner E. Gerabek described an early-twentieth-century German custom in which toothache sufferers hung worms around their necks, "waited until the little worms had died, then . . . threw the worms into an oven and prayed the Lord's Prayer."

Meanwhile, back in the eighteenth century, the influential dentist Pierre Fauchard slammed his boot down on tooth worms with the 1728 publication of his magnum opus, *Le Chirurgien Dentiste, ou Traité des Dents* (translated as *The Surgeon Dentist, or Dental Treatise*). In it, he promoted scientific technique over quackery and folklore—choosing to perform his own experiments, rather than falling back on the teaching of a Roman physician who had been holding his breath for fifteen hundred years. After repeated unsuccessful efforts to find worms in the mouths of his patients, Fauchard wrote that while worms might be consumed accidentally, they were clearly not responsible for toothaches or tooth decay.

Helping to complete the transition of tooth worms into the realm of folklore was the American dentist Willoughby Miller, whose late-nineteenth-century research led him to develop the chemoparasitic hypothesis of tooth decay in 1890.* With it, he demonstrated that it wasn't the burrowing behavior of tooth worms that was responsible for cavities but the action of oral bacteria. Their metabolisms produced an acid byproduct that was literally bound in place against the tooth surface by accumulations of plaque (otherwise known as tartar or calculus). When this stuff hardened, Miller hypothesized, the acid would dissolve a tooth's protective enamel covering. All of this, he assured his readers, would take place without any assistance from worms.

But what about modern-day folks—folks who generally take care of their teeth and who would hardly ever allow plaque to build up to the point where it resembles a hockey player's pale-yellow mouth guard?

Having spent *many* long hours with my fingers locked viselike onto the armrests of dozens of dentist chairs across Long Island, I became even more uneasy during my annual tartar-extraction fest when my dental hygienist, Theresa Felton, asked me if I wanted to come in the following week to see some cavity-causing oral bacteria up close. I had mentioned that I was particularly interested in the habitat shared by these microbes and the way they divvy up the resources they find there. At that point, and just as I was getting used to the novel concept of visiting a dental office willingly, my dentist entered the room. Felton quickly gave him the lowdown.

"Great," Dr. Sheren replied before turning to me. "She can take the samples from you."

---

* Also known as the acid dissolution theory.

I gave the suggestion careful consideration for approximately two seconds, then figured out a time when my mouth and I could come in.

As the day approached, I found myself becoming more and more enthusiastic about what was about to take place. I knew that long before the building of New York City's gigantic Jacob K. Javits Convention Center, my own mouth had served as a popular convention hall for a large percentage of the plaque-inhabiting bacteria in the northeastern United States. What better place, then, I reasoned, to investigate the real-life descendants of the legendary tooth worm?

When the time came, Felton and I decided to sample two very different oral ecosystems: one above the gumline (home to the surface-dwelling supragingival microbes) and the other below the gumline (realm of the Morlock-equivalent subgingival bacteria). (See figure 38.)

Under the heading of "Do Not Try This at Home," Felton wielded a sharpened-at-both-ends metal instrument called a scaler, using it to wrestle a tiny bit of tartar from the cramped space between my lower incisors.

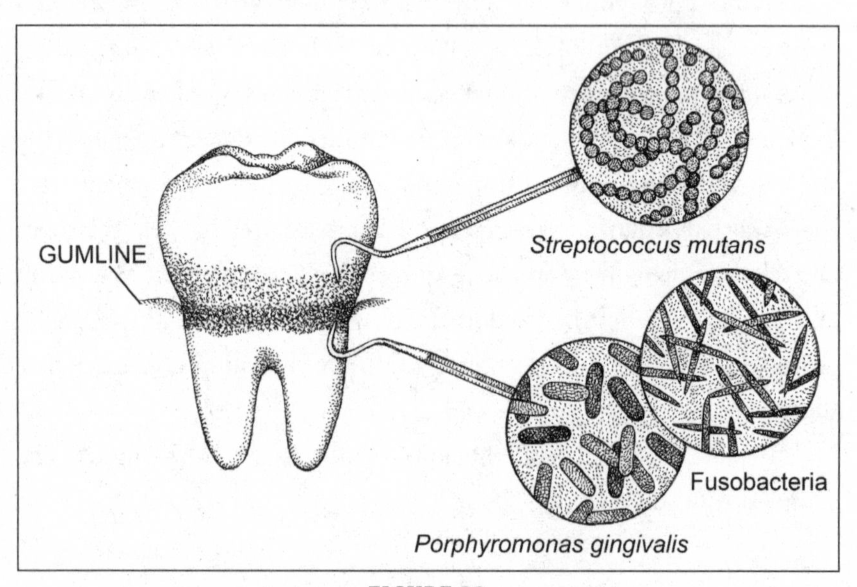

FIGURE 38

After applying some of my glop to a microscope slide, she added a glass coverslip. Next, Felton prepared a second slide, this one after briefly probing around *below* the gumline of one of my lower molars.

She set up the first slide on a compound microscope, this one equipped with a nifty video camera that would enable us to view the magnified action on a monitor. I had seen nightmarish photos of the so-called bacterial consortia inhabiting supragingival plaque—dense assemblages of tiny dots, squiggles, and rods. Admittedly twitchy, given my history of tooth rot and gum problems, I expected the worst.

Surprisingly, though, this slide looked rather barren—vast empty spaces with occasional tiny round dots.

The dental hygienist tentatively identified these as *Streptococcus mutans*, the most common of the several hundred bacterial species living topside of the gumline.* These are also the microbes whose quest for adenosine triphosphate, through the process of fermentation, results in most cavity-related tooth problems.

The chemical breakdown of fermentable carbohydrates (like sugar), which releases ATP, also produces lactic acid.† The resulting drop in pH to around 5.5 (anything below 7 is considered acidic) in the plaque-encrusted area where the bacteria are doing their thing leads to the acid-assisted breakdown of tooth enamel and dentin located below it, a process known as demineralization.‡

The "problem" we now ran into was that I had been to the dentist

* A close relative and cohabitant, *Streptococcus sobrinus*, is indistinguishable at this level of examination.

† In the alcoholic beverage industry, fermentation, in which yeast break down glucose, is basically the same process, but instead of lactic acid, it produces ethyl alcohol (ethanol).

‡ The pH scale (which reads from 1 to 14) is an indicator of how acidic or alkaline a substance is. Since the scale is logarithmic, the numbers are inverse indicators of the concentration of hydrogen ions (H+). In other words, low pH means a lot of hydrogen ions, and vice versa. And since acid strength varies directly with H+ concentration, the lower the pH is below 7, the stronger the acid. The higher the number above 7, the more alkaline (or basic) the substance is.

the week before. That visit had kicked off with the amiable dental hygienist applying some serious elbow grease in an apparently successful effort to dislodge a year's worth of plaque from the very same set of pearly yellows she had just finished probing. "You're not exciting," she said as we scanned the sparsely populated slidescape. "Which is a good thing."

I decided to remain silent, especially given the fact that she still had a trayful of sharp instruments within arm's reach.

Next, Felton secured the slide containing the subgingival plaque onto the microscope stage. She'd obtained this sample from an area she had described to me the previous week as a particularly troublesome spot. It was also one I'd been directed to spend extra time cleaning with an electric toothbrush after a good power blasting with my trusty Waterpik.

As the second slide came into focus, it immediately became clear that the term "troublesome spot" had been an understatement. The entire field of view was alive with movement, and I was immediately reminded of a big budget battle scene in the last of the *Lord of the Rings* flicks. There appeared to be thousands of creatures of seemingly every size and shape tearing around and bashing into each other, and some put-upon-looking white blood cells. But instead of in Middle-Earth, the action was taking place against a backdrop of spongy-looking brown material.

Felton pointed out that this was the hardened biofilm known in microbial circles as "home." Prominent among the frantic horde, primarily because of their significantly larger size, were rod-shaped microbes called fusobacteria—a group of pathogens noted for their involvement in gum disease, particularly the inflammation and irritation of the gums known as gingivitis.

Like *S. mutans*, fusobacteria do not require oxygen to survive (making them anaerobic organisms), but they can use it if it happens to be

lying around (which further classifies them as facultative anaerobes).*
This dual metabolic system is a good thing if you happen to live in an environment where there's always a chance that there might be some residual oxygen around, perhaps trapped in periodontal pockets, spaces under the gumline where bacteria tend to congregate in massive numbers. As fusobacteria infect the surrounding gums, the resulting deep infection (i.e., periodontitis) can lead to bone and tooth loss.

There are also bacteria for which even small amounts of oxygen can be deadly—the aptly named obligate anaerobes. One such below-the-gumline oxygen hater is *Porphyromonas gingivalis*. (See figure 38.) Not only does this species contribute to gum disease, but it has recently been implicated as a possible causative agent in the development of Alzheimer's disease. Studies have shown that *P. gingivalis* cannot survive by itself in an environment where even small amounts of oxygen is present. The facultatively anaerobic fusobacteria, which share their subgingival habitat with *P. gingivalis*, will burn up any oxygen if it's around. This makes the surrounding environment more conducive to the survival and reproduction of *P. gingivalis*—much to the misery of the mouth owner.

Another difference between *S. mutans* and fusobacteria is the way that they secure ATP. Unlike *S. mutans*, which breaks down sugar to release ATP, fusobacteria are proteolytic, meaning that they degrade (i.e., cut up) proteins and their chemical building blocks, amino acids. The byproducts of this chemical reaction cause the surrounding pH to rise toward a more neutral reading of around 7.

One of the reasons that pH is vitally important is because the thousands of chemical reactions that take place in the body and elsewhere can only occur within their own specific range of pH. The proteins classified

* Aerobic organisms (or aerobes) require oxygen to survive. They include plants, fungi, some bacteria, and all animals.

as enzymes (due to their ability to accelerate chemical reactions) serve as the perfect example. For instance, the enzyme known as salivary amylase, which starts the chemical digestion of starches in the mouth, works best at a pH of between 6 and 7. You can test this yourself by chewing on a piece of bread (or a potato, if you prefer) for twenty seconds or so. The sweet taste that develops as you chew results from the starch (a carbohydrate) being broken down into sugar by salivary amylase.

Once the saliva containing the salivary amylase is swallowed and reaches the highly acidic stomach (pH of 1.5–3.5), that enzyme is denatured. In other words, it loses its shape and stops working. As a result, starch digestion does not pick up again until partially digested food leaves the stomach and enters the duodenum, the upper portion of the small intestine. There, thanks to additions of alkaline secretions from the pancreas into the small intestine, the pH rises to around 6. Significantly, the pancreas also adds pancreatic amylase, and because this enzyme works just fine at pH 6, it completes the job of starch breakdown.

Chemical reactions like these happen throughout the body, making it essential that the body constantly regulates and maintains pH. It does so in part with the aid of chemical receptors located on the inner walls of large blood vessels, like the carotid artery. These chemoreceptors monitor the blood for stuff like pH and send signals to the brain when pH adjustments in either direction are required.*

For our little bit of the story, it's important to know that studies have shown that in the mouth, neutral or slightly alkaline environments are better for obligate anaerobes like *P. gingivalis*, while the supragingival

---

* Like pH, temperature is also a limiting factor for chemical reactions, which is one reason it's dangerous for the body to get too hot (high fever) or too cold (hypothermia) for too long. For example, like changes in pH, high temperatures can also denature enzymes. But like chemoreceptors located internally that detect pH levels, thermoreceptors in the skin alert the brain when body temperatures get too high or too low. Among the responses would be sweating and shivering, respectively.

microbes like *S. mutans* do best when things are more acidic. These different pH environments are produced and maintained by the microbes that thrive in these respective conditions, and they also serve as a way to exclude outsiders that can't tolerate that particular pH.

Leaving the topic on something of a foul note: The byproducts of protein and amino-acid breakdown by oral bacteria include sulfur compounds, ammonia, and a pair of chemicals called indole and skatole—better known for giving feces its distinctive smell. Having these metabolic products in the mouth results in halitosis (bad breath), a condition that can be exacerbated by poor dental hygiene or eating certain foods (like cheese and citrus fruit).

In some ways, our role as amateur dental hygienists is simply to keep the bacterial bash from getting out of hand. And although this is inevitably a losing battle, hopefully we can take some small comfort knowing that none of the partygoers are tooth worms.

[17]

Wisdom Teeth, Baby Teeth, and the Tooth Fairy

I just thought the tooth fairy was a very creepy concept as a kid, "Put your tooth under the pillow." I was like "Why does someone want my teeth?"

—Guillermo del Toro

The concept of vestigial dentition—otherwise known as "teeth that no longer serve a purpose"—is one that is likely familiar to many readers. That's because our rearmost upper and lower molars also go by a more notorious name: wisdom teeth. (See figure 27, page 147.) These particular molars usually (though not always) make their unwelcome presence known when we are somewhere between the ages of eighteen and twenty-five years old.

Wisdom teeth are also a reminder that our ancient ancestors had a longer, more protruding jaw than we do. Their last-to-arrive molars presumably worked just fine. Unfortunately for modern humans, while our jaws shortened over evolutionary time, the part of our genetic blueprint responsible for tooth number and, to a lesser extent, tooth size did not change, and now there simply isn't enough room for these latecomers, especially in individuals with smaller jaws. Instead, the crowns of the wisdom tooth crowns erupt from the gums incompletely or not at all, conditions referred to as partially or fully impacted. A small percentage of people experience no health-related problems with their wisdom teeth, but for most of us, they crowd or damage other teeth. Impacted wisdom teeth also significantly increase the risk of infection, and in doing so they often produce swelling, pain, and misery.

Found throughout the animal kingdom, vestigial organs also include the sightless and reduced eyes found in blind cave fish, wings in flightless birds, and functionless pelvic (and even hind-limb) bones in whales and pythons. In humans, our coccyx (or tailbone) is a prime example. Then there are the barely-there molars found in the three vampire bat species. Like human wisdom teeth, these tiny nubs serve no purpose in the blood feeders, though they belong to the far more popular pain-free variety.

Researchers believe that wisdom teeth were fully functional when our longer- and sturdier-jawed ancestors were busy chewing and grinding down tough, uncooked meat and plant matter. Serving to increase the crushing ability of the jaws, an extra pair of upper and lower molars was a valuable adaptation in early *Homo sapiens* and their ancient ancestors, especially given the severe wear and tear they would have been subject to over the life of the individual.

Then, somewhere between fifteen thousand and ten thousand years ago, as humans learned to grow and prepare softer food, our jaws gradually became shorter and more delicately built—with smaller jaw muscles requiring bones with less surface area for their attachment. One possible reason for the jaw shortening in *Homo sapiens* may relate to the energy expense required to maintain the hefty muscle and bone tissue previously on hand to process those tougher food items. According to the authors of a recent study, "With tools, food processing, cooking, and agriculture, modern humans are liberated from lengthy bouts of daily mastication." Once softer, processed food became available, natural selection would have worked to gradually reduce the size of the overbuilt jawbones and the bulky muscles that moved them.

On a related note, according to Tanya Smith, while impacted wisdom teeth currently plague one out of every four people living today, the problem is more common for some than it is for others. In her book, *The Tales Teeth Tell*, Smith mentioned that Native Americans, Asians, and Middle Easterners commonly lack an extra set of molars. She and her former graduate student Kate Carter hypothesized that this condition may be a relatively recent evolutionary response to the inherent danger of wisdom-tooth-related health risks. In other words, before the advent of modern dental care, those individuals who did not have an extra set of molars were *more* likely to live to reproductive age (and pass on that genetic trait) than those who were in danger of dying from infections resulting from impacted wisdom teeth.

I asked Smith if she believed that wisdom teeth would eventually go the way of the human tail—in other words, would they disappear?

"Not unless we stop providing medical care for their removal," she replied. "Typically, natural selection acts on things that make it hard to reproduce and leave healthy offspring, so if their presence isn't fatal, wisdom teeth will likely persist in some people."

Finally, while teaching students about evolution, I have very often used vestigial organs like wisdom teeth as evidence that life on our planet was not the result of "intelligent design" but a complex set of evolutionary processes that often turned out far-from-perfect results. I mean, why give us a tailbone whose only function seems to be to break if we land on it or even sit down too hard?

As usual, Charles Darwin summed up the concept nicely by comparing vestigial (a.k.a. rudimentary) organs to the silent letters in a word:

> Rudimentary organs, by whatever steps they may have been degraded into their present useless condition, are the record of a former state of things, and have been retained solely through the

power of inheritance . . . [They] may be compared with the letters in a word, still retained in the spelling, but become useless in the pronunciation, but which serve as a clue for its derivation.

BABY TEETH

Human teeth begin forming during the first trimester of development, when epithelial cells, a type of cell that lines body cavities and covers surfaces, migrate into the developing jaws. These cells stimulate an embryonic connective tissue known as mesenchyme, to form bell-shaped clusters called tooth buds. From their positions within the developing jaws, the tooth buds grow progressively larger, gradually taking on the shape of incisors, canines, premolars, or molars.

In a step that emphasizes the importance of a nutritious diet for pregnant mothers, necessary minerals like calcium, phosphorus, and fluoride are carried to the site of the tooth buds by the circulatory system. With the arrival of these mineral building blocks, development continues at an accelerated rate. Dentin is laid down in a region that will eventually become the tooth crown and the roots that will secure the teeth to the jaw. Other cells use the incoming minerals to deposit enamel at the surface of the tooth crown as it develops. (All of this action takes place within the jawbones.)

As the first crown erupts through the gums, the jaw tissue surrounding the tooth solidifies, secured to the roots by cementum and cord-like periodontal ligaments.*

Although many vertebrates exhibit unlimited tooth replacement, most mammals have but two sets, and these exhibit significant differences in number, size, shape, and composition. For example, while most

* Readers interested in a more complete description of tooth embryology (and developmental anomalies) can find it at the NIH's National Library of Medicine, https://www.ncbi.nlm.nih.gov/books/NBK560515.

humans have thirty-two permanent teeth, there are only twenty deciduous precursors (also known as baby teeth or milk teeth). These erupt from the gums between the ages of 6 months and 33 months in *approximately* the following order: lower and upper central incisors (5–12 months), lateral incisors (9–16 months), first molars (13–19 months), canines (16–23 months), second molars (23–33 months).

There are also dissimilarities in tooth structure that make perfect sense for anyone who has dealt with baby teeth and the individuals who carry them around. For example, the layer of enamel covering these teeth is relatively thinner than it is in adult teeth. And although this can make baby teeth look whiter than their adult counterparts, it also explains why they are more susceptible to cavities. Deciduous teeth also have roots, whose presence may come as a surprise, since most of us don't get to see them. That's because those tooth roots eventually dissolve and are resorbed, with calcium and phosphorus recovered by the circulatory system. That doesn't happen with adult teeth.

Similar to their emergence (incisors first, molars last), the shedding of deciduous teeth is a sequential event, in this case partially mediated by cells called odontoclasts. These jumbo-sized cells form when macrophages merge. Macrophages are a type of white blood cell whose day job is to hunt down and engulf pathogens like bacteria. Sometimes, though, they fuse together into odontoclasts, recognizable by the presence of multiple nuclei (a reminder that these cells were formerly a group of soloists, each with a single nucleus). Odontoclasts function by secreting enzymes and acids that break down bone tissue. When activated outside the roots of deciduous teeth, their bone-dissolving action results in loosening of the teeth. This makes it easier for them to be shed and replaced by the adult teeth already growing toward them from above or below.

In addition to dissolving the dentin and cementum found in deciduous tooth roots, similarly multinucleated cells, known as osteoclasts, serve on the cleanup team that eliminates bone fragments after a fracture. Their action also functions to smooth over uneven sections of bone that commonly occur during the process of bone repair.

TV commercial voice: But wait, there's more. When blood calcium

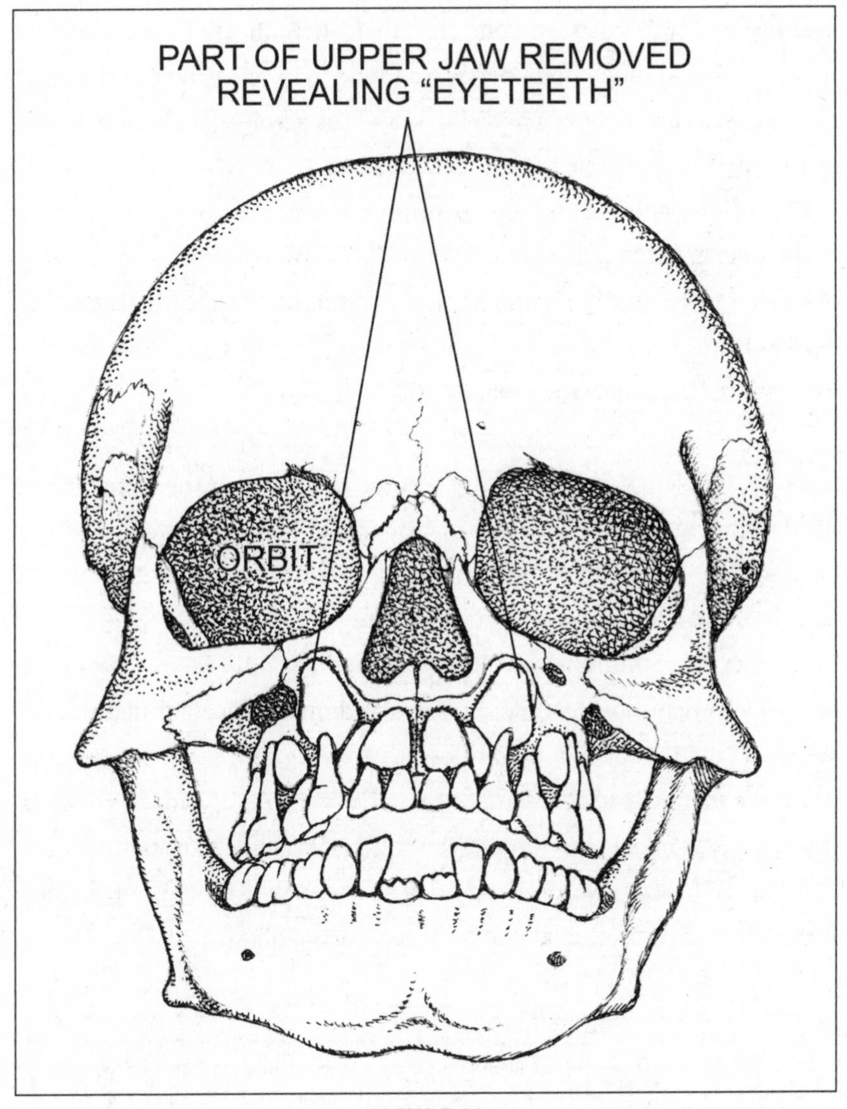

FIGURE 39

levels are low, osteoclasts break down the bone matrix to free up calcium for use by the body. In healthy individuals, this bone resorption is balanced by the deposition of new bone tissue—with the two processes referred to as bone remodeling. When an imbalance occurs, bones become filled with empty spaces, causing them to resemble a sponge and leading to osteoporosis—a condition that often affects long bones like the femur.* Too much bone deposition (due to excess growth hormone) can lead to a rare medical condition known as acromegaly, which results in enlarged bones of the hands, feet, and face.

Before the upper deciduous canines are shed, the on-deck position of the unerupted upper adult canines—just above the deciduous ones—has given them an interesting alternative moniker: eyeteeth. The name derives from the fact that they sit just below the orbits, the bony cavities in the skull that house the eyeballs. (See figure 39.)

I SPOKE TO Nancy Simmons, the curator-in-charge in the Department of Mammalogy at the American Museum of Natural History, about a curious type of deciduous teeth present in the creatures that she and I have been studying for most of our careers: bats.

Unlike me, Simmons has closely studied bat milk teeth. "Bats have diphyodont dentition just like humans and other placental mammals," she said, "but the form of the milk teeth are very different." She went on to tell me that instead of resembling smaller versions of bats' adult teeth, the baby teeth in bats are "small, recurved hooks with sharp points." (See figure 40.) This backward-facing hook shape enables a baby bat to hook on securely to its mother as she carries it during flight.

---

* A reduction in the bone mass in the narrow region located between the femoral shaft and the ball-like femoral head is a common cause of broken hips in those suffering from osteoporosis. Here, a fall or even just the body weight of an individual can initiate a fracture in the structurally compromised region.

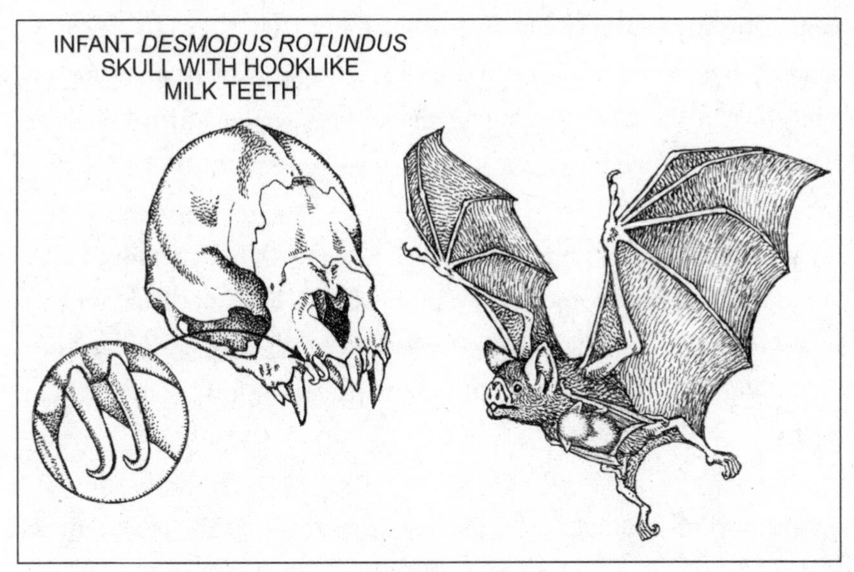

FIGURE 40

"Like Velcro, the baby teeth don't puncture the mother's skin," Simmons explained, "but they do allow for a tight hold on the mother's nipple. Bat wings are really modified arms and hands, so a mother bat cannot hold its baby in its arms and fly at the same time. To stay safe, a baby bat needs to hold on tight to mom when she flies, and hooklike baby teeth apparently evolved to help with this."

THE TOOTH FAIRY

Before leaving baby teeth, it is absolutely necessary to address an issue of related and great importance: namely, the tooth fairy.

Compared to Santa Claus (a.k.a. Sinterklaas, Saint Nicholas, Sint Nicolaas, Kris Kringle, etc.), who was born in around 280 CE in Patara, in what is now Turkey, and the Easter Bunny, who hopped onto a ship bound for the United States with some German immigrants in the eighteenth century, the tooth fairy is a relatively new arrival. According to Garrett Williams, self-described "lawn-game enthusiast" and distinguished tooth fairy expert, press coverage of the tooth fairy's exploits began on September 27, 1908, in a short piece by newspaper reader Lillian Brown, published in the *Chicago Daily Tribune*, in a section titled "Practical Housekeeper's Own Page":

> Many a refractory child will allow a loose tooth to be removed if he knows about the tooth fairy. If he takes his little tooth and puts it under a pillow when he goes to bed the tooth fairy will come in the night and take it away, and in its place will leave some little gift.*

The next mention of the tooth fairy, according to Williams, occurred in the May 14, 1911, edition of Louisville, Kentucky's *Courier-Journal.* In an article, fifteen-year-old Nellie May Hans recounts placing a tooth under her pillow when she was six. "After I had fallen asleep the 'Tooth Fairy' came over to my bed, picked up the tooth and said to me, 'Nellie May, would you like to go to Fairyland with me?'"

Evidence of Nellie May's answer can be found in the title of the piece, "My Visit to Fairyland."

The tooth fairy gained even more acclaim after playwright Esther Watkins Arnold penned a children's play aptly titled *The Tooth Fairy* in 1927.

The rest, as they say, is history.

More recently, the folks at Delta Dental have been calculating the average value of a lost tooth in the United States. When they started doing so in 1998, the payoff was around $1.30 per tooth. By 2023, their Original Tooth Fairy Poll found that the amount had soared to $6.23.

---

* I'm guessing that Brown's use of the term "refractory child" did not refer to one who was resistant to decomposition by heat, pressure, or chemical attack, but rather to a kid who was stubborn or unmanageable.

Kids may want to temper their expectations, though, since Garrett Williams reported that the tooth fairy "tends to be more generous in the Northeast and stingier in the South and West." Additionally, loose-toothed tots may want to keep an eye on the stock market, since the average haul from the tooth fairy typically tracks the ups and downs of the S&P 500.

[18]

To Infinity and Beyond . . . the Dental Chair

The power to control our species' genetic future is awesome and terrifying. Deciding how to handle it may be the biggest challenge we have ever faced.

—Jennifer Doudna, CRISPR pioneer

Toothless creatures notwithstanding, the importance of teeth in the natural world and in the lives of humans, past and present, cannot be overemphasized. Exciting research into teeth, from the perspectives of both natural history and dental medicine, where new therapies and technologies are already in use, is becoming more widespread. In addition to improving patient outcomes, these innovations are making trips to the dentist less frequent and, yes, even less traumatic.

In one example, many dentists are now using computer-assisted design (CAD) and computer-assisted manufacturing (CAM) to produce crowns—tooth-shaped caps placed over the drilled-down remains of damaged teeth. This new method eliminates time-consuming steps like the production of molds and temporary crowns, as well as the necessity of having off-site dental labs manufacture the permanent crowns. Instead, digitized images of the original tooth are sent to a 3D printer, which builds the crown on-site. The material used is also new: a ceramic known as zirconium oxide, which is stronger than porcelain. It also resists developing sharpened edges that can damage the tongue or the inside of the mouth. This technology is beginning to transform the production of dentures and other prosthetic appliances from one requiring multiple trips to the dentist into the always-preferable one-shot affair.

Looking to an only slightly more distant horizon, researchers at the University of Nottingham and Harvard University are seeking to make composite tooth fillings obsolete. They hope to replace them with so-called regenerative fillings, which will stimulate stem cells to repair and regrow regions damaged by tooth worms.* According to Adam Celiz, a research associate at Nottingham, they "have designed synthetic biomaterials that can be used similarly to dental fillings but can be placed in direct contact with pulp tissue to stimulate the native stem cell population for repair and regeneration of pulp tissue and the surrounding dentin."

The hope is that regenerative fillings will one day relegate the term "root canal" to the same dustbin of dental terminology that currently holds "pelicans" and "bow drills."

In another important development, medical researchers are currently investigating the now-suspected link between the common subgingival bacterium *Porphyromonas gingivalis* and Alzheimer's disease. Not only does this progressive and incurable brain disorder consistently rank in the top ten causes of death in the United States, but it is responsible for an enormous emotional and financial toll on people with Alzheimer's disease, along with their family and friends. Currently, there are more than six million Americans with Alzheimer's, a number projected to reach approximately thirteen million by 2050.

P. gingivalis is a main cause of the severe gum infection known as periodontitis. The harmful action of *P. gingivalis* involves the release of

---

* Just checking to see if you were paying attention.

toxic substances called gingipains. Like the previously discussed metalloproteases found in viper venom, these are enzymes that break down protein. In the mouth, gingipains destroy cells and tissues within the gums. This can trigger inflammation—a protective response by the body that can be harmful when it lasts too long, as in chronic inflammation, or when it occurs in healthy tissues.

In an extensive 2019 study, physician and scientist Stephen Dominy and twenty-five coauthors showed that levels of *P. gingivalis* were significantly higher in the brains of Alzheimer's disease patients than they were in control individuals (i.e., people without the disease). A key characteristic of Alzheimer's disease is the production of amyloid plaques, hard clumps of misfolded protein that accumulate in the brain around neurons (especially those in areas concerned with memory), disrupting their function. Dominy and his collaborators demonstrated that in test mice exposed to *P. gingivalis* orally, the bacterium was able to invade the brain. There, *P. gingivalis* not only produced increased levels of a component found in amyloid plaques, but it also "had detrimental effects on tau, a protein needed for normal neuronal function." The study authors proposed that *P. gingivalis* and the gingipains it releases play key roles in the development of Alzheimer's disease.

Dominy and others are now working to develop so-called gingipain inhibitors. In their initial experiments on mice, these molecules (which must be small enough to pass through the blood-brain barrier) reduced neuroinflammation, blocked the neural damage caused by gingipains, and reduced the bacterial load (i.e., the amount of *P. gingivalis* present). As of March 2022, clinical trial results were mixed, although one compound (a bacterial protease inhibitor) significantly slowed cognitive decline in a subset of study participants, namely those who had detectable levels of *P. gingivalis* in their saliva.

But while the jury remains out on the effectiveness of gingipain inhibitors, there is growing evidence that *P. gingivalis* is a causal factor in Alzheimer's disease. Given the staggering number of victims and the likelihood that the healthcare system and the economy will incur even greater stresses, the relationship between Alzheimer's disease and *P. gingivalis* warrants closer scrutiny and increased funding for research. Stay tuned.

I HOPE BY now you have a greater appreciation for how teeth are used by their owners. Researchers also have a long record of drawing on teeth, the most common of all vertebrate fossils, to explore the history of life on this planet, both past and present. So I asked Tanya Smith for an example of fossil teeth that were instrumental in our current knowledge about a group of organisms.

"The fossil record of the great apes—chimpanzees, orangutans, and gorillas—over the past several million years is almost entirely understood from teeth," she told me. "There are no known skeletons from ancient chimpanzees and gorillas, despite them being in Africa for seven to ten million years. In the case of the Asian orangutan, thousands of fossil teeth have been found, but only one skeleton has been recovered, despite them having lived in this region for millions of years."

In the twenty-first century, there is still a relative scarcity of so-called postcranial evidence (i.e., from the neck down), though not nearly to the extent faced by earlier researchers. Even so, teeth still play an important role in modern research.

We have already seen how climate change research, based on the presence of carbon isotopes, has contributed to the understanding of how ancient herbivores, like horses, evolved as transitions occurred in the types of plants available to them. Likewise, evidence for the presence of C_3 and C_4 isotopes has provided important information on the changing diets of "hominins," a term used to describe modern humans and our extinct bipedal relatives.

Studies on the carbon-isotope content of fossil hominin teeth from eastern Africa have shown that between 4.4 and 4 mya, early hominins had C_3 plant-based diets. You'll recall that C_3 plants are generally soft and leafy while C_4 plants are grasses and grains. Around 3.5 mya, several hominin species, including *Australopithecus afarensis*, made a transition to a broader diet of C_3 and C_4 plants. These C_4 plants were not actually new arrivals, but had instead been around for a million years before they became menu items. Once this dietary transition occurred, though, the inclusion of C_4 grasses and grains would become a hallmark of the expanding diets of all subsequent hominins including *Homo sapiens*.

Though the ancestors of modern humans evolved in Africa, eventually some of these groups migrated out—in stages. And reminiscent of how the simple straight-line evolution of horses has been discredited, the out-of-Africa migration of hominins did not occur in an orderly, linear fashion. Instead, this oversimplified version of the human story has been replaced by an understanding that the actual events were far more complex, with migrant groups overlapping, competing, and even interbreeding.

One such group was the Neanderthals, who lived in Eurasia from roughly 230,000 years ago until around 40,000 years ago. This means that they were already present when modern humans migrated out of Africa around 45,000 years ago. Beginning in 2008, human bone fragments and three teeth were discovered in Denisova Cave in Siberia. The fossils were thought to be around 75,000 years old. Genetic analysis of DNA extracted from the fossils ultimately indicated that the so-called Denisovans were neither Neanderthal nor modern human, though they were more closely related to the former and had interbred with them. Although Denisovan fossils are extremely rare, the discovery of additional teeth in a cave on the Tibetan plateau expanded their known range into Tibet but no farther east. This begged the question of how Denisovan DNA wound up in modern-day Indigenous Southeast Asians and Pacific Islanders.

In a 2022 study, University of Illinois paleoanthropologist Laura Shackelford and her colleagues recounted the discovery of a single molar in a cave in northern Laos, a landlocked country in Southeast Asia. After comparing numerous features of the tooth's crown to those of other ancient humans, the researchers concluded that the best match for the tooth was that it came from a Denisovan. And since the molar had no roots, they knew it was a deciduous tooth, either a first or second lower molar. Further analysis suggested that it belonged to a young female, since an enamel protein associated with the male-determining Y chromosome was missing.

Knowing that they had to be careful making claims based on a single tooth, the researchers were still able to offer evidence that would potentially further expand the known range of the Denisovans. More importantly, this bit of physical evidence could help explain how Denisovan DNA occurs in modern humans from Southeast Asia to Australia. Researchers believe that the Denisovans went extinct around thirty thousand years ago, but apparently not before they spent some quality time with modern humans in places like Laos.

For more definitive information on the diets and behavior of past humans, some scientists are looking for evidence in the form of microremains locked within a substance that has long plagued those individuals with poor dental hygiene. That material is dental calculus (a.k.a. tartar or plaque). Its color can vary from individual to individual and is sometimes dependent on the color of whatever a person happened to be eating or drinking—with shades of white, gray, black, and even green possible.

Besides its unsightly appearance, dental calculus can result in health problems because it provides a sturdy haven to harmful bacteria. But this mineralized matrix also traps material like hair, pollen, starch grains, and phytoliths (hardened deposits of silica found in many plant taxa). These microremains are important to researchers, not only because they can be used to identify the specific plants or animals that produced them, but also because they were clearly part of an ancient diet. You can get some idea of what we're talking about here if you consume a three-course meal, use a water-powered flosser, and then look at the bottom of your bathroom sink.

Now, let your mind wander back to the PW era (pre-Waterpik).

When I was a child, Neanderthals were generally depicted as shambling losers who hung around in caves waiting around to be eaten by saber-toothed cats. Much of this had to do with reconstructions of the initial Neanderthal fossils. For reasons that ranged from prejudice to working with a specimen that was seriously arthritic, nineteenth-century anthropologists reassembled the first Neanderthal skeletons to resemble stooped-over apes, and the late-nineteenth-century and early-twentieth-century artwork depicting Neanderthals reflected this.

We now know that these humans were anything but losers. Having moved out of Africa, they became widespread across Eurasia for almost two hundred thousand years. They were highly intelligent, with a successful culture and a toolbox full of complex stone implements that helped them survive some extremely harsh environmental conditions. Then, somewhere around forty thousand years ago, they disappeared.

Although climate change is sometimes invoked to explain the Neanderthal extinction, the more favored hypotheses stress the arrival of modern humans, who had begun migrating to Europe from Africa only a few thousand years before the Neanderthal abruptly vanished. The popular take on what occurred usually focuses on specific aspects of technology or behavior that gave the newly arrived modern humans an edge. The story never ends well for the Neanderthals who, depending on whom you ask, were either quickly driven to extinction, assimilated into the modern human gene pool, or both. As for how modern humans might have pulled this off, the list of potential competitive advantages includes deadlier weapons, better clothing (to cope with cold climates), larger group sizes, and a longer life expectancy.

Another hypothesis suggests that a more diverse diet, which incorporated fish and plants, allowed modern humans to take advantage of limited resources in ways that the Neanderthals hadn't. Until the last few decades, Neanderthals were generally depicted as subsisting primarily on meat, and only recently has there been any evidence that they also consumed plants. And although Neanderthals were known to have used fire, there was little evidence that they used it to cook their food.

To investigate Neanderthal diets further, archaeologist Amanda Henry and her colleagues analyzed dental calculus scraped from the teeth of Neanderthal skeletons. Her sample came from specimens from both cold and warm climates (Belgium and Iraq). In both samples, the researchers found evidence that a variety of plant foods had been consumed, including date palms, legumes, and grass seeds. They also determined that grass seed starches recovered from the calculus showed damage indicative of having been cooked. Their results suggested a more varied diet for Neanderthals as well as a higher degree of sophistication in the preparation of plant foods than had been previously recorded.

Dental calculus contents can also provide information on diet when local conditions make alternative forms of data collection difficult. In a 2019 study, a team including Henry and University of Helsinki doctoral student Tytti Juhola examined thirty-two dental calculus samples taken from a previously excavated Iron Age cemetery in the town of Eura, in southwestern Finland, where the approximate age of the corpses buried there had been dated to 600–900 CE.

The problem researchers face in studying boreal (northern, cold-climate, forested) environments like this one, is that the long-term deposition of plant material, especially pine needles, turns the soil and sediments acidic. This acidity causes the rapid decomposition of any organic material, especially if it contains calcium, a primary component of bones. This generally makes it more difficult to study the history of the vertebrates, including humans, that inhabited these regions, even those who lived there relatively recently. As in this case, though, the extrahard nature of teeth sometimes allows evidence to be preserved.

After scraping tiny bits of dental calculus from these previously exhumed human remains, the researchers used light and polarized microscopy to analyze them and determine their contents. Though the sample size was small, they uncovered plant fibers (flax, hemp, and nettles), which may have been from garments or food; mammal hairs, from sheep and mountain hares (*Lepus timidus*); and feathers, including one from some kind of waterfowl. The researchers suggested that the feathers may have come from a pillow or might have been inhaled while someone was plucking a bird.

Though the resulting evidence added only a few pieces to the puzzle of what life was like for this population of Iron Age Finlanders, the study of dental calculus is a developing field and has already become an important tool for researchers.

MUCH HAS BEEN learned from the study of vertebrate teeth, but there is a need for caution. Though by no means exclusive to the sciences, researchers can get into trouble when they come to conclusions after asking too much from the evidence on hand—in this instance, using teeth to make sweeping statements about major groups of animals or their behavior.

Tanya Smith has warned that while tooth shape can be a strong indicator of diet in animals like herbivores and carnivores, "it's not possible to simply look at the teeth of fossil hominins and know which specific foods they ate."

She used conodonts as a good example of misleading conclusions drawn from dental remains. In this instance, what some thought to be plant parts turned out to be evidence of a diverse and highly successful group of jawless fish likely exhibiting a variety of lifestyles.

For over a hundred years, though, these ancient and mysterious aquatic creatures were known only from the tiny toothlike structures they left behind. Conodonts were first described in 1856, by Latvian embryologist Christian Pander (1794–1865), who believed that they were fish teeth. Some early paleontologists believed that the fossils weren't actually teeth but were instead the remains of plants, marine worms (e.g., chaetognaths), earthworms, mollusks, and even tiny organisms that *looked* like teeth.

It wasn't until an articulated fossil was unearthed in Scotland that researchers determined the toothlike structures it contained (now known as conodont elements) were actually from jawless, eel-shaped creatures. The picture became even more clear when similar whole-body fossils were discovered in South Africa.

Conodonts are now believed to be distant relatives of hagfish. A major difference, though, is that conodonts were small, with average body lengths of between 1 centimeter (0.4 inch) to 40 centimeters (16 inches), and their "teeth" were really small, ranging from 0.2 millimeters (.008 inches) to 5 millimeters (0.2 inches) in length (see figure 41).

Researchers now believe that conodonts made up approximately fifty families and over fifteen hundred species. We also know that beyond

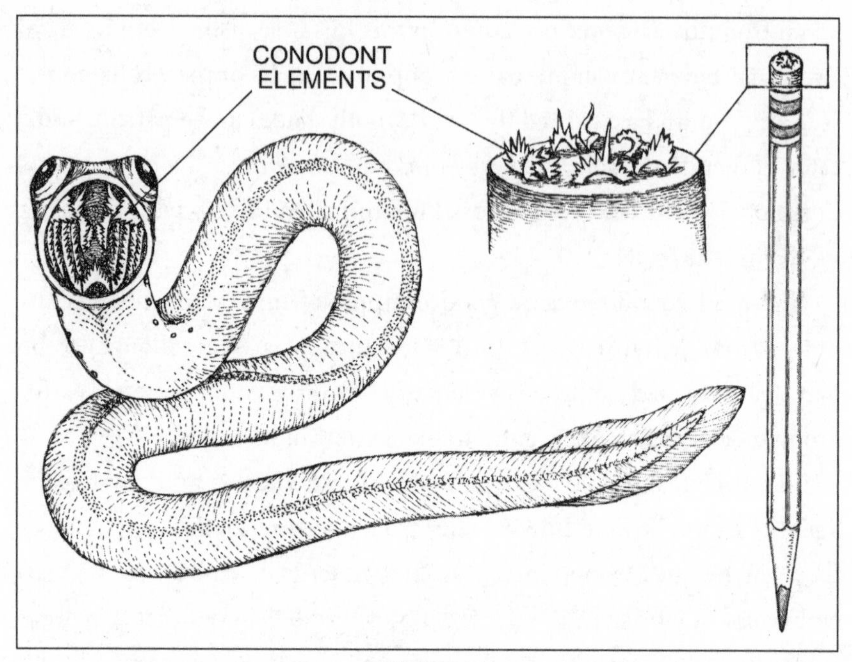

FIGURE 41

their diversity, conodonts were around for a remarkably long stretch of time: from the Late Cambrian period until the end of the Triassic, a span of roughly three hundred million years.

Examined in the context of new evidence, a 130-year-old mystery had been solved. And while there is no evidence that the tiny toothlike structures found in the jawless conodonts ever evolved into vertebrate teeth, these fossils are still providing paleontologists with information on the age of ancient rocks, the boundaries between geologic time periods, and the reconstruction of ancient oceans and climates.

Scientists are often asked to justify their research in response to questions like, "So vampire bats are cool and all, but how does studying them help my grandmother live longer?" Oftentimes, what the public might perceive as relevance comes later and in unexpected ways.

In vampire bats, it was the discovery of desmokinase, also known as *Desmodus* salivary plasminogen activator (DSPA). This blood-clot-busting enzyme was isolated from vampire bat saliva and has become an effective agent in the treatment of strokes.

Similarly, the study of teeth as time capsules, providing archaeologists with information on the early life experiences of ancient people, may provoke similar questions about relevance from the public or, worse yet, from folks in charge of distributing grant money. But the concepts behind using teeth as records of a long-dead individual's exposure to stressors like disease or environmental factors may soon be taking a new and important turn—one that relates to the possible prevention of mental health disorders.

We have already seen how pioneering work by researchers like Tanya Smith and others have used the tree-ring-like incremental growth lines of dentin and enamel in teeth to determine precisely when stressful events like birth, injury, infection, disease, and exposure to harmful chemicals occurred. Now, Boston University clinical psychology PhD student Kathryn A. Davis and her colleagues are investigating the potential creation of a program that would use information obtained from teeth on early life adversity (such as food deprivation or exposure to environmental toxins) to predict the increased likelihood of conditions like depression, anxiety, autism, and schizophrenia in youths.

In addition to an established track record of studies utilizing fossil teeth to preserve ancient life experiences, Davis and her colleagues cite additional advantages for using freshly shed primary teeth as "biomarkers" of early life trauma. These advantages include considerations like objectivity and the accessibility and inexpensive nature of the toothy samples. Davis also emphasized the fact that since primary teeth grow and fall out over a dozen years or so, analysis of those individual teeth would give those examining them "a specific time window," from prenatal development to the onset of the teen years, during which stressors could be detected.*

Of course, questions, and indeed hurdles, remain, though Davis and her colleagues are the first to admit it. Some are logistical. Others are social or cultural in nature. For example, who would collect the teeth destined for examination? Parents? That could get dicey. Dentists are an obvious answer, but who would they pass those teeth on to? In other words, who would do the analysis? And how would interventions and preventive care be initiated if the analysis of a child's tooth set off alarm bells? Clearly, new ties between dentists, pediatricians, mental health providers, and social services would need to be established.

Although the list of concerns to be addressed is a long one, the end result could be a childhood screening program with the potential to dramatically reduce the occurrence of mental health disorders during middle childhood or later—and such a program would be truly transformative.

MEANWHILE, PING WU and his fellow researchers at the University of Southern California and elsewhere are hoping to learn useful medical information from alligators. Specifically, they studied their polyphyodont-style tooth replacement which can grow upward of fifty sets of replacement teeth. Unlike most creatures with so-called unlimited tooth replacement, alligator teeth reside in alveoli that resemble those of mammals in structure. Their teeth also show additional resemblances to the mammalian tooth condition, notably the steps involved in their replacement. In brief, Wu and his colleagues were able to painstakingly map out the complex sequence of developmental steps that occur as lost teeth are replaced over the course of an alligator's life—which can

* Since the second molars undergo development up to sixteen years of age, their examination could potentially extend the period of early adversity detection into the teen years.

extend for over seventy years. They utilized tissue from embryos, hatchlings, and juvenile gators, plus computer models, determining that tooth replacement in alligators involves three distinct steps: preinitiation, initiation, and growth.

A key to the entire process, though, is a region of tissue called the dental lamina. In alligators, when a tooth is lost, the dental lamina activates latent stem cells as well as the cascade of chemical signals that initiate tooth formation. Humans also have dental laminas (a.k.a. laminae) involved in tooth replacement, but they become inactive after our permanent teeth come in. Wu and his colleagues believe that what they've learned about tooth regeneration in alligators may one day allow us to restart the dormant tooth-formation process in adult humans.

Taking a different path, some researchers in Japan believe that humans actually have a third set of tooth buds—the cell clusters that develop into individual teeth. This would be huge news, since humans were thought to have but two sets. Sayaka Mishima and his colleagues at Kyoto University and elsewhere are attempting to activate this hypothesized third batch of tooth buds in cases where they might be needed.

Their initial studies on mice sought to treat a human condition known as inherited tooth agenesis, in which up to six adult teeth fail to develop. After determining the cause to be several possible gene mutations, the researchers now claim that they can reverse the results of these mutations (at least in mice), leading to what they term "rescued tooth development." Clinical trials of a topically applied medication that they hope will do just the same for humans are planned for 2024.

If successful, studies like these could usher in a new age in dental medicine—eliminating the need for dentures in instances when teeth were accidentally lost, were removed due to infection, or never emerged in the first place.

This work illustrates yet another of the myriad connections between humans and the creatures that share this planet with us—here similarities in the dental biology of alligators, mice, and humans. And while not specifically addressing the public's concerns that animal studies might not prolong Grandma's life, these and future avenues of research at least have the potential to keep her smiling with all of her own teeth.

Epilogue

How Science Works

November 2023

"Isn't that the strangest skull you've ever seen?

The question was from my friend, artist and scientific illustrator Patricia Wynne, whose drawings grace this book. Though we'd both been associated with the American Museum of Natural History for decades, it was still a special treat to wander the halls before the doors opened and the great crowds poured in. Today, we were standing in front of a display case in the Hall of Advanced Mammals—and we'd come to look at teeth.

I stepped in closer, peering through the glass and feeling slightly embarrassed that I'd probably walked past this particular exhibit a hundred times without ever giving it a whole lot of thought. At first, I concentrated on the four-foot-long upper portion of the fossil skull, which was mounted just below eye level. I knew that it was an elephant relative, not just because of two forward-facing sulci that had once held a pair of small tusks, but because of the condition of the two large molars on each side of the upper jaw. The dead giveaway was the fact that the rearmost of these teeth appeared wear-free, whereas those sitting in front of them had been ground into nubs. As you'll recall, elephants and their relatives are unusual in that their molars are replaced not from below, as in most mammals, but conveyor-belt-style, by fresh reinforcements from the rear. It was one of many things I had never known about elephants until I started looking into their mouths.

But beyond all the new stuff I had learned about teeth, I'd also developed a greater awareness that many of the tales told by teeth begin with a prologue in which the hero has died. For example, regardless of whether the teeth and jaws in question are millions of years old (as in this particular specimen) or had come from a modern elephant whose skull resides in a museum collection, the tooth-replacement process had been halted suddenly, frozen in time by the death of the individual. Now, we can see the process midstream, when the old teeth had been worn down but the new had not yet replaced them—something that would be pretty much impossible to see in a living specimen. In death, though, each specimen has taken on a new role: to educate me, as well as a bunch of other folks, about how this particular dental process works and how it differs from the condition seen in most mammals.

Studying teeth has also given me a renewed appreciation for the variation that can exist within the class Mammalia—a group I'd mostly thought about in terms of characteristics they shared (their organs, for example). While you might need an expert to differentiate a pig heart from a human heart, when it comes to teeth, the differences are hard to miss. Here, under the banner of tooth variation, the lower jaw that Patricia had guided me to was clearly a showstopper. Yes, it held a pair of molars on each side, just like the upper jaw, but that was pretty much where the similarities ended. For starters, this mandible was significantly longer than its upper partner, giving the skull a two-foot underbite. It was in this extended and toothless region of the lower jaw, just in front of the molars, that things took a sharp turn toward the bizarre. After dipping downward, the right and left lower jawbones converged briefly, then flared out dramatically, forming a wide, upward-facing scoop. Then, from the jaw's scalloped frontmost edge emerged a pair of enormous incisors. Flattened on both their upper and lower surfaces (dorsoventrally flattened), their total width spanned what I estimated to be about eighteen inches (see figure 42A).

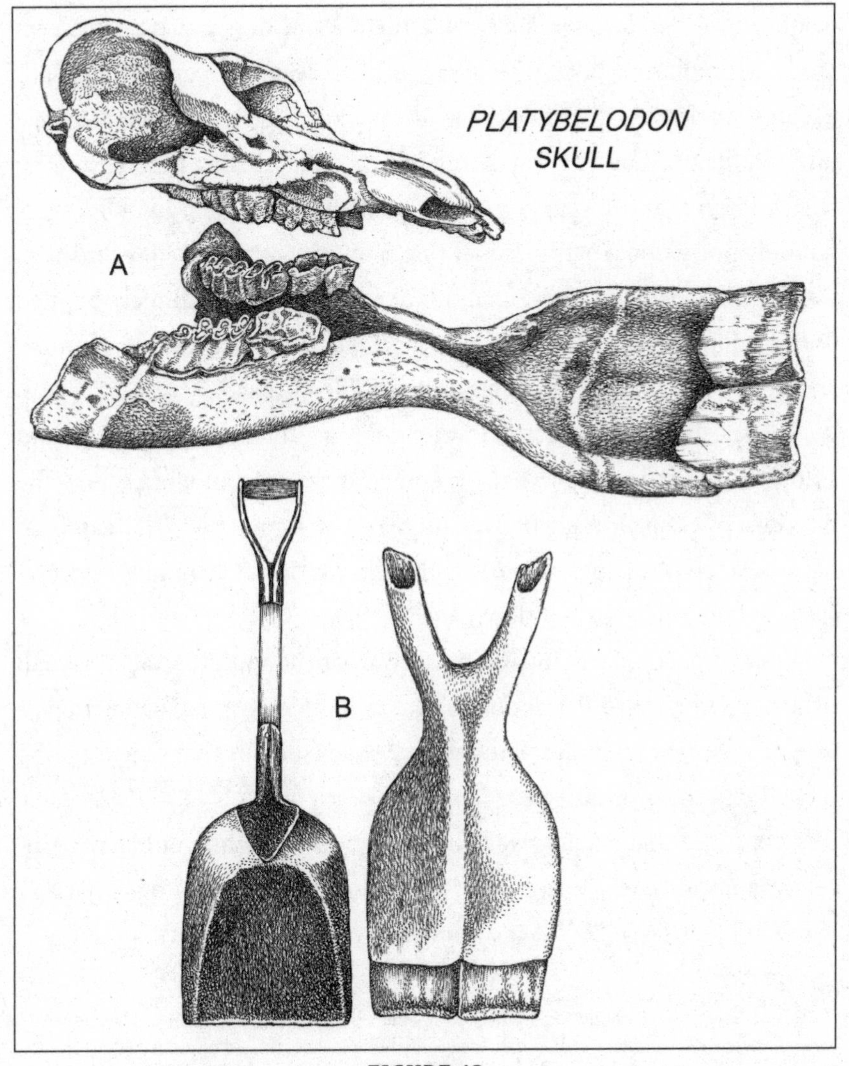

FIGURE 42

"*Well*, isn't that the strangest skull you've ever seen?" Patricia repeated.

I stared at the lower jaw a moment longer, unable to shake the image of a rather hefty shovel. "It's right up there," I replied, though in truth I couldn't come up with anything to top it.

On the museum label inside the display case, I read that the skull belonged to *Platybelodon grangeri*, a member of the extinct pachyderm family Gomphotheriidae. The gomphotheres were a diverse group with a nearly worldwide distribution, which peaked during the Late Miocene (around five million years ago) and went extinct during the Pleistocene (2.58 million to 11,500 years ago). Unlike today's elephants, gomphotheres not only possessed a pair of upper tusks, but they had two lower tusks as well. In *Platybelodon*, these mandibular tusks would definitely be filed under *S* for "strange" but also for "spectacular." Their unique anatomy resulted in the creature's common name: shovel-tusker.*

The specimen Patricia and I were looking at had been one of around a dozen collected in the late 1920s and early 1930s from what is now the province of Gansu, in northern China, by the famous AMNH explorer, naturalist, mammalogist, and fossil collector Roy Chapman Andrews (1884–1960) and his expedition team.

I read a particularly intriguing section of the museum tag out loud: "The beveled ends of *Platybelodon*'s tusks have led some paleontologists to speculate that the animal might have sharpened its tusks by dragging them along the ground."

I could see that Patricia was skeptical about how this could have conceivably been done—so I decided to demonstrate, jutting out my chin, Don Corleone–style (or, as it's known in the trade, pronating the lower

---

* There are generally thought to have been four genera of gomphotheres referred to as shovel-tuskers (including *Amebelodon* in North America), though none sported mouthparts anywhere near as extreme as *Platybelodon*.

jaw), then bending at the waist and shuffling forward, directing my open mouth through an imaginary Miocene landscape, which likely looked nothing like the tiled fourth floor of the museum.

"Great, Bill," Patricia said, laughing. "You just got a mouthful of dirt and rocks."

Unfazed, I spit out my mouthful of imaginary slag and reversed direction, dragging my imaginary shovel tusks on the ground as I backed up.

"No way. Too unwieldy," was Patricia's assessment.

Out of the corner of my eye, I noticed that a security guard had wandered over to see what was going on. I straightened up, smiled, and offered him a brief wave, which he returned with an uptilt of his head. Then, seeing that Patricia and I were wearing museum ID badges, he moved on without comment, presumably seeking the company of some nonscientist types.

As for what *Platybelodon* was doing with those jumbo incisors (when it wasn't making a fool of itself, that is), the museum tag told us that it "may have been used to root out vegetation, especially around lakes and marshes."

We continued our meandering route through the storied halls of the AMNH, passing below the towering skeletons of *Platybelodon*'s mammoth relatives, several of which were actual mammoths, and performing our own version of a dental examination. At one stop, Patricia spun a toothy tale of male camels murdering each other in a paddock in back of the Smithsonian sometime during the post–Civil War nineteenth century. We took a slight detour to see her favorite amphibian (*Mastodonsaurus giganteus*), a 210-million-year-old giant frog relative with what looked to be a pair of upward-facing tusks protruding through its near-five-foot-long skull. Our tour ended in front of an ancient shark relative that had mosaiclike teeth that reminded me of the parrotfish teeth I'd written about and Patricia had drawn.

BACK HOME, I couldn't get *Platybelodon*'s weird lower incisors out of my mind. Something just didn't seem right. Investigating further, I discovered the origin of the museum tag information on "shovel tusks."

In 1932, AMNH vertebrate paleontologist (and eugenicist) Henry Fairfield Osborn and coauthor Walter Granger, also an AMNH vertebrate paleontologist, wrote: "With few exceptions, *Platybelodon* has been found in direct association with quantities of fresh-water bi-valves [clams and their pals], suggesting that this region was at that time the shore-line of a lake of considerable extent. *Platybelodon*, with its broad flattened lower tusks, evidently found the region to its liking and very likely used its shovel-shaped jaw to scoop up succulent water plants." The authors were somewhat less than subtle regarding their proposed tusk-as-shovel association, including a photograph of *Platybelodon*'s lower jaw sitting upright and next to a coal shovel (see figure 42B).

The image struck a chord, and the creature was often depicted in books, and even museum displays, dutifully wading through marshes and dredging water weeds, wielding its mouthparts like an enormous spork.

Then in 1992, paleontologist David Lambert proposed something different. He noted the wear pattern of *Platybelodon*'s oversized incisors and used theoretical models of tusk wear to propose that just because something *looks* like a shovel doesn't always mean that it *is* a shovel. Instead, Lambert suggested that the foreshortened upper jaw meant that the creature had a trunk very much like that of modern elephants, not the abbreviated version generally depicted for *Platybelodon*. He hypothesized that this appendage was used to grasp branches, which were then run up against the beveled edge of the lower incisors—much like we might use a saw—freeing the branches for consumption.

Although Lambert didn't provide much in the way of physical evidence, I thought it was a neat hypothesis.

In 2016, the ancient-elephant tale received yet another twist. Gina Semprebon, professor of biology at Bay Path University, and her colleagues assessed the diet of *Platybelodon* through two separate analyses. In the first, they examined its molars, to determine the microscopic wear patterns found on their enamel surface. This technique, which had been in use for several decades, provided the researchers with what they called "a snapshot of dietary behavior" during a short period just before the tooth owner died. They used statistical analysis to compare their results from *Platybelodon* to a large database consisting of tooth wear patterns in mammals with known diets—including the three species of extant elephants.

They also examined seven mandibular tusks for signs of gouges and microscopic scars—doing so to test the hypothesis that *Platybelodon*'s shovel tusks were indeed an appropriate name. What they found, though, was far more consistent with a diet of leaves and twigs, with *Platybelodon*'s molar wear patterns resembling those of the extant forest elephant (*Loxodonta cyclotis*). Their results also indicated that the finely patterned scratches on the mandibular tusks did not support their use as shovels to scoop aquatic or semiaquatic plants—which would have presumably shown more serious wear from rocks and grit-laden water weeds. Instead, Semprebon and her colleagues found, as Lambert had hypothesized, it was more likely that the mandibular tusks had been used to strip vegetation and bark off shrubs and trees, similar to the way extant elephants do with their tusks (see figure 43).

Platybelodon's lower jaw is a fascinating tooth-related adaptation, but the course science took to understand this tusk's function is just as intriguing and perhaps more important: Proposed by scientific heavyweights nearly a century ago, one explanation gains widespread acceptance. Some sixty years later, another researcher uses new techniques to

FIGURE 43

come up with a very different hypothesis, but the question remains an open one. A quarter century after that, scientists revisit the very same question—this time with an array of still newer tools and techniques, unavailable to any of the past investigators—sending the initial hypothesis into something akin to retirement.

This, in a nutshell, is the scientific process. And as these things go, this most recent interpretation will almost certainly undergo further change down the road as future researchers employ even newer tools and techniques to strengthen or modify the existing hypothesis or to propose a new hypothesis of their own. There is never a last word in science.

As for teeth, we've now seen that much of the diversity and evolutionary success of the vertebrates can be attributed to their presence. As new specimens are discovered and old specimens reexamined, teeth, the most common and important vertebrate fossils, will undoubtably continue to provide us with new insights into the human condition, natural history, and the world around us.

Notes

Introduction

3 **According to the World Health Organization** "Snakebite Envenoming," World Health Organization, https://www.who.int/news-room/fact-sheets/detail/snakebite-envenoming.

4 **teeth as tools** Katherine J. Wu, "Like Neanderthals, Early Modern Humans Used Their Teeth as Tools," *NOVA*, November 27, 2019, https://www.pbs.org/wgbh/nova/article/neanderthals-early-modern-humans-teeth-tools.

6 **London bill of mortality** Bernhard Wolf Weinberger, *An Introduction to the History of Dentistry in America* (St. Louis, MO: C. V. Mosby, 1948), 12.

6 **nightmarish photos** Brett Kelman and Anna Werner, "This Dental Device Was Sold to Fix Patients' Jaws. Lawsuits Claim It Wrecked Their Teeth," *CBS Mornings*, March 2, 2023, ttps://www.cbsnews.com/news/agga-dental-device-lawsuits-teeth-damage.

Chapter 1: Vampire Bats Don't Suck

11 **Gonzalo Fernando de Oviedo** David E. Brown, *Vampiro: The Vampire Bat in Fact and Fantasy* (Silver City, NM: High-Lonesome Books, 1994), 13.

16 **Zoologist Dennis Turner** Dennis C. Turner, *The Vampire Bat: A Field Study in Behavior and Ecology* (Baltimore: Johns Hopkins Press, 1975), 65.

16 **wound-feeding hypothesis** M. B. Fenton, "Wounds and the Origin of Blood-Feeding in Bats," *Biological Journal of the Linnean Society* 47, no. 2 (October 1992): 161–71, https://doi.org/10.1111/j.1095-8312.1992.tb00662.x.

16 **arboreal-feeding hypothesis** W. A. Schutt Jr., "The Chiropteran Hindlimb Morphology and the Origin of Blood Feeding in Bats" in *Bat Biology and Conservation*, ed. Thomas H. Kunz and Paul A. Racey (Washington, DC: Smithsonian Institute Press, 1998), 157–68.

19 **"the pinnacle of specialization"** Arthur M. Greenhall, *Natural History of Vampire Bats* (New York: Routledge, 1988), 112.

19 **blood feeders of every ilk** Bill Schutt, *Dark Banquet: Blood and the Curious Lives of Blood-Feeding Creatures* (New York: Harmony Books, 2008).

20 **lack or lose the enamel coating** C. J. Phillips and B. Steinberg, "Histological and Scanning Electron Microscopic Studies of Tooth Structure and Thegosis in the Common Vampire Bat, Desmodus rotundus," *Occasional Papers, The Museum, Texas Tech University* 42 (September 10, 1976): 1–12.

21 **questioned the absence of enamel** John W. Hermanson and Gerald C. Carter, "Vampire Bats," in *Phyllostomid Bats*, ed. Theodore H. Fleming, Liliana M. Dávalos, and Marco A. R. Mello (Chicago: University of Chicago Press, 2020), 257–72.

Chapter 2: Candirus: Be Careful Where You Go

25 **"preoccupied with doing harm"** Paul Le Cointe, *L'Amazonie Brésilienne; Le pays—Ses habitants, Ses ressources*[;] *Notes et statistiques jusqu'en 1920*, vol. 2 (Paris: Augustin Challamel, 1922) in Stephen Spotte, *Candiru: Life and Legend of the Bloodsucking Catfishes* (Berkeley, CA: Creative Arts Book, 2002), 25.

26 **"another fish which is dangerous to man"** From the preface of Johann de [von] Spix and Louis Agassiz's text *Selecta genera et species piscium quos in itinere per Brasiliam annis MDCCCXVII–MDCCCXX*, published between 1829 and 1831, in Spotte, *Candiru*, 12.

26 **horror stories gained traction** E. W. Gudger, *The Candiru* (New York: Paul B. Hoeber, 1930).

31 **"completely negative results"** Spotte, *Candiru*, 143.

31 **"Fatal Stream Hypothesis"** Spotte, 154–55.

32 **"entered in the boy's urethral canal"** Spotte, 212.

32 **"*darted out of the water*"** Spotte, 213.

33 **"The trip to the urethra"** Spotte, 217.

33 **Spotte did find this problematic** Spotte, 214.

35 **"being struck by lightning"** Schutt, *Dark Banquet*, 281.

Chapter 3: Horses: Long in the Tooth

36 **"Rather than simplify"** Bruce J. MacFaddden, *Fossil Horses: Systematics, Paleobiology, and Evolution of the Family Equidae* (Cambridge: Cambridge University Press, 1992), 325.

38 **the Bone Wars** Mark Jaffe, *The Gilded Dinosaur: The Fossil War Between E. D. Cope and O. C. Marsh and the Rise of American Science* (New York: Three Rivers Press, 2001), 432.

40 **a variety of vegetation** MacFaddden, *Fossil Horses*, 236.

41 **"more diverse in their body sizes"** B. J. MacFadden, "Fossil Horses—Evidence for Evolution," *Science* 307, no. 5716 (2005): 1728–30, https://www.science.org/doi/10.1126/science.1105458.

42 **10 percent of dry weight** S. Kumar and R. Elbaum, "Estimation of Silica Cell Silicification Level in Grass Leaves Using *in situ* Charring Method," *Bio-Protocol* 7, no. 22 (November 20, 2017): e2607, https://www.ncbi.nlm.nih.gov/pmc/articles/PMC8438387.

43 **this stony material** R. A. Stirton, "Observations on Evolutionary Rates in Hypsodonty," Evolution 1 (1947): 32–41, https://doi.org/10.2307/2405401; G. Semprebon, F. Rivals, and C. Janis, "The Role of Grass vs. Exogenous Abrasives in the Paleodietary Patterns of North American Ungulates," *Frontiers in Ecology and Evolution* 7 (March 2019): 1–23, https://doi.org/10.3389/fevo.2019.00065.

50 **Contrary to Cope's "bigger is better" mantra** MacFadden, "Fossil Horses," 1730.

Chapter 4: The Tusked and the Tuskless

55 **the arrangement allowed for continuous growth** M. R. Whitney et al., "The Evolution of the Synapsid Tusk: Insights from Dicynodont Therapsid Tusk Histology," *Proceedings of the Royal Society B* 288, no. 1961 (October 27, 2021): 8, https://doi.org/10.1098/rspb.2021.1670.

60 **"It proved no better than blacklead"** George Best, "Frobisher: Second Voyage (1577)," in *The North-West and North-East Passages, 1576–1611*, ed. Philip F. Alexander (Cambridge: Cambridge University Press, 1915), 23.

60 **"the sea-unicorn"** Best, 23–24.

61 **examined 131 narwal skulls** M. Nweeia et al., "Vestigial Tooth Anatomy and Tusk Nomenclature for *Monodon monoceros*," *Anatomical Record* 295 (2012):1006–16, https://doi.org/10.1002/ar.22449.

63 **tiny channels in the cementum** F. C. Eichmiller and D. H. Pashley, "The Extraordinary Narwhal Tooth," in *Narwhal: Revealing an Arctic Legend*, 58–65.

64 **this sensory system evolved** M. T. Nweeia et al. "Sensory Ability in the Narwhal Teeth Organ System," *Anatomical Record* 297 (2014): 599–617, https://doi.org/10.1002/ar.22886.

64 **easier access to breathing holes** Eichmiller and Pashley, 64–65.

66 **Evidence from a drone video** Fisheries and Oceans Canada, "Canadian Scientific Footage Confirms Use of Narwhal Tusk," news release, May 12, 2017, https://www.canada.ca/en/fisheries-oceans/news/2017/05/canadian_scientificfootageconfirmsuseofnarwhaltusk.html. .

67 **"Whales lack the physiological properties"** Peter Bondo, "The Narwhal's Tusk Reveals Its Past Living Conditions," March 11, 2021, Phys.org, https://phys.org/news/2021-03-narwhal-tusk-reveals-conditions.html

67 **In places like northern Canada** C. M. Pirkle, G. Muckle, and M. Lemire, "Managing Mercury Exposure in Northern Canadian Communities," *Canadian Medical Association Journal* 188, no. 14 (October 4, 2016): 1015–23), https://www.ncbi.nlm.nih.gov/pmc/articles/PMC5047817.

68 **The number of teeth that manatees have** "Manatee Adaptations: The Head," Journey North, University of Wisconsin-Madison Arboretum, https://journeynorth.org/tm/manatee/AdaptationsHead.html.

70 **elephants were slaughtered by poachers** S. C. Campbell-Staton et al., "Ivory Poaching and the Rapid Evolution of Tusklessness in African Elephants," *Science* 374, no. 6566 (October 21, 2021): 483–87, https://www.science.org/doi/10.1126/science.abe7389.

70 **selection for tusklessness arose** Campbell-Staton et al., 483.

70 **percentage of tuskless females in those areas increased** H. Jachmann, P. S. M. Berry, and H. Imae, "Tusklessness in African Elephants: A Future Trend," *African Journal of Ecology* 33, no. 3 (1995): 230–35, https://doi.org/10.1111/j.1365-2028.1995.tb00800.x.

70 **a mutation on the X chromosome** Campbell-Staton et al., 486.

71 **the corresponding genes in female *humans*** G. M. Hobson et al., "A Large X-Chromosomal Deletion Is Associated with Microphthalmia with Linear Skin Defects (MLS) and Amelogenesis Imperfecta (XAI)," *American Journal of Medical Genetics* 149A (2009): 1698–1705, https://doi.org/10.1002/ajmg.a.32968.

71 **the incidence of tusklessness** E. J. Raubenheimer, "Development of the Tush and Tusk and Tusklessness in the African Elephant (*Loxodonta africana*)," *Koedoe* 43, no. 2 (2000): 57–64, https://doi.org/10.4102/koedoe.v43i2.199.

71 **"Tuskless males do not occur"** Campbell-Staton et al., 486.

71 **male Asian elephants** "Asian Elephant," Smithsonian's National Zoo & Conservation Biology Institute, https://nationalzoo.si.edu/animals/asian-elephant.

71 **humans interacting with Asian elephants** Raubenheimer, 62–63.

Chapter 5: Fangs a Lot

73 **"it promptly bit me"** Christie Wilcox, "Bite from the Past: New Study on Boomslang Venom Provides Insights into the Death of Renowned Herpetologist Karl Schmidt," *Discover*, January 30, 2017, https://www.discovermagazine.com/planet-earth/bite-from-the-past-new-study-on-boomslang-venom-provides-insights-into-the-death-of-renowned-herpetologist-karl-schmidt.

77 **"bleeding of mucus membranes in the mouth"** Kaushik Patowary, "Karl Patterson Schmidt: The Herpetologist Who Documented His Own Death for Science," Amusing Planet, January 9, 2020, https://www.amusingplanet.com/2020/01/karl-patterson-schmidt-herpetologist.html.

77 **"anal plate was undivided"** Elizabeth Shockman and Alexa Lim, "He Documented His Own Death by Snakebite Instead of Going to the Hospital," Science Friday, *The World*, November 10, 2015, https://theworld.org/stories/2015-11-10/he-documented-his-own-death-snakebite-instead-going-hospital.

78 **an antivenin designed for use against rattlesnakes** D. Pla et al., "What Killed Karl Patterson Schmidt? Combined Venom Gland Transcriptomic, Venomic and Antivenomic Analysis of the South African Green Tree Snake (the Boomslang), *Dispholidus typus*," *Biochimica et Biophysica Acta (BBA)—General Subjects* 1861, no. 4 (April 2017): 814–23, https://doi.org/10.1016/j.bbagen.2017.01.020.

79 **The ingredients in a venom cocktail** T. M. Abd El-Aziz, A. Garcia Soares, and J. D. Stockand, "Snake Venoms in Drug Discovery: Valuable Therapeutic Tools for Life Saving," *Toxins* 11, no. 10 (2019): 564, https://doi.org/10.3390/toxins11100564.

81 **"most known toxins have been described only incompletely"** Abd El-Aziz et al., "Snake Venoms."

000 **treat conditions like arthritis** A. Gomes, "Snake Venom—An Anti Arthritis Natural Product," *Al Ameen Journal of Medical Sciences* 3 (2010): 179.

83 **the mechanics of rattlesnake bites** K. V. Kardong and V. L. Bels, "Rattlesnake Strike Behavior: Kinematics," *Journal of Experimental Biology* 201 (1998): 837–50, https://doi.org/10.1242/jeb.201.6.837.

83 **"extend, contact, release/retract"** Kardong and Bels, 841.

85 **evolved to immobilize the prey as quickly as possible** D. Cundall, "Viper Fangs: Functional Limitations of Extreme Teeth," *Physiological and Biochemical Zoology* 82, no. 1 (2009): 63–79, https://doi.org/10.1086/594380.

85 **regulate the amount of venom they inject** M. E. Peterson, "Snake Bite: Pit Vipers," *Clinical Techniques in Small Animal Practice* 21 (2006): 174–82, https://doi.org/10.1053/j.ctsap.2006.10.008.

85 **permanent muscle damage** H. F. Williams et al., "Mechanisms Underpinning the Permanent Muscle Damage Induced by Snake Venom Metalloprotease," *PLoS Neglected Tropical Diseases* 13, no. 1 (2019): e0007041, https://doi.org/10.1371/journal.pntd.0007041.

86 **In some vipers and elapids** S. P. Mackessy and L. M. Baxter, "Bioweapons Synthesis and Storage: The Venom Gland of Front-Fanged Snakes," *Zoologischer Anzeiger: A Journal of Comparative Zoology* 245, nos. 3–4 (2006): 147–59, https://doi.org/10.1016/j.jcz.2006.01.003.

86 **antibodies to venom components** R. Straight, J. L. Glenn, and C. C. Snyder, "Antivenom Activity of Rattlesnake Blood Plasma," *Nature* 261 (1976): 259–60, https://doi.org/10.1038/261259a0.

86 **modified receptors on the membrane surface** D. Servent et al., "How Do Snake Curaremimetic Toxins Discriminate Between Nicotinic Acetylcholine Receptor Subtypes," *Toxicology Letter* 102–103 (1998): 199–203, https://doi.org/10.1016/S0378-4274(98)00307-5.

87 **development of the drug captopril** Abd El-Aziz et al.

Chapter 6: Poop on the Beach: The Good Kind

90 **crystals of fluorapatite** Glenn Roberts Jr., "X-Rays Reveal the Biting Truth About Parrotfish Teeth," Berkeley Lab, November 15, 2017, https://newscenter.lbl.gov/2017/11/15/xrays-reveal-biting-truth-about-parrotfish-teeth; M. A. Marcus et al., "Parrotfish Teeth: Stiff Biominerals Whose Microstructure Makes Them Tough and Abrasion-Resistant to Bite Stony Corals," *ACS Nano* 11, no. 12 (2017): 11856–65, https://doi.org/10.1021/acsnano.7b05044.

91 **Surgeonfish take a gentler . . . approach** Tye Kindinger, "'A Better Chance for Resilience': Using Hungry Fish to Conserve Coral Reefs," NOAA Fisheries, April 17, 2022, https://www.fisheries.noaa.gov/feature-story/better-chance-resilience-using-hungry-fish-conserve-coral-reefs.

91 **one study conducted in the Bahamas** M. A. Albins and M. A. Hixon, "Invasive Indo-Pacific Lionfish *Pterois volitans* Reduce Recruitment of Atlantic Coral-Reef Fishes," *Marine Ecology Progress Series* 367 (2008): 233–38, https://doi.org/10.3354/meps07620.

91 **"one of the greatest threats of this century"** Invasive Lionfish Web Portal, http://lionfish.gcfi.org/index.php.

92 **the agency's recommendation** "Impacts of Invasive Lionfish," NOAA Fisheries, July 29, 2022, https://www.fisheries.noaa.gov/southeast/ecosystems/impacts-invasive-lionfish.

93 **toxin-producing glandular tissue** S. Raju, "Three New Species of the Genus *Monognathus* and Leptocephali of the Order Saccopharyngiformes," *Fishery Bulletin* 72, no. 2 (1974): 547–62, https://spo.nmfs.noaa.gov/fishery-bulletin-journal/722.

93 **how its fang is employed during the bite** R. J. Harris and R. A. Jenner, "Evolutionary Ecology of Fish Venom: Adaptations and Consequences of Evolving a Venom System," *Toxins (Basel)* 11, no. 2 (2019): 60, https://doi.org/10.3390/toxins11020060.

94 **"loose aggregations** G. S. Losey, "Predation Protection in the Poison-Fang Blenny, *Meiacanthus atrodorsalis*, and Its Mimics, *Ecsenius bicolor* and *Runula laudandus* (Blenniidae)," *Pacific Science* 26 (April 1972): 131–32.

94 **a 2017 paper on . . . blennies** N. R. Casewell et al., "The Evolution of Fangs, Venom, and Mimicry Systems in Blenny Fishes," *Current Biology* 27, no. 8 (April 24, 2017): 1184–91, https://doi.org/10.1016/j.cub.2017.02.067.

96 **"The typical reaction"** Losey, 135.

98 **viceroys are as unpalatable as monarchs** D. B. Ritland and L. P. Brower, "The Viceroy Butterfly Is Not a Batesian Mimic," *Nature* 350 (1991): 497–98, https://doi.org/10.1038/350497a0.

Chapter 7: Shrews: Tiny in Size, Major in Attitude

103 **why venomous shrews aren't affected by their own venom** Rachel Sheremeta Pepling, "The Stunning Saliva of Shrews," *Chemical & Engineering News* 82, no. 42 (October 13, 2004), https://cen.acs.org/articles/82/i42/Stunning-Saliva-Shrews.html.

Chapter 8: Bite This!

105 **"I knew how to use a shovel,"** Andrew Grant, "5 Questions: From Fossils in Rocks to Stadium Rock," *Discover*, June 1, 2009, https://www.discovermagazine.com/planet-earth/5-questions-from-fossils-in-rocks-to-stadium-rock.

114 **a twenty-one-foot great white shark** S. Wroe et al., "Three-Dimensional Computer Analysis of White Shark Jaw Mechanics: How Hard Can a Great White Bite?" *Journal of Zoology* 276, no. 4 (2008): 336–42, https://doi.org/10.1111/j.1469-7998.2008.00494.x.

Chapter 9: By the Teeth of Their Skin

121 **"The tinkerer gives his materials unexpected functions"** F. Jacob, "Evolution and Tinkering," *Science* 196, no. 4295 (June 10, 1977): 1164, https://www.science.org /doi/10.1126/science.860134.

125 **attach itself to his own neck** Discovery Channel Southeast Asia, "Sea Lamprey," *River Monsters*, YouTube video, May 14, 2014, https://www.youtube.com/watch?v=uVh4UF-3H9A.

133 **the "outside-in hypothesis"** Alfred S. Romer, *Vertebrate Paleontology* (Chicago: University of Chicago Press, 1936), 492; Malcolm Jollie, "Some Implications of the Acceptance of a Delamination Principle," in *Current Problems of Lower Vertebrate Phylogeny*, ed. Tor Ørvig (Stockholm: Almqvist and Wiksell, 1968): 89–108; W. Reif, "Evolution of Dermal Skeleton and Dentition in Vertebrates: The Odontode Regulation Theory," in *Evolutionary Biology* 15, ed. Max K. Hecht (New York: Plenum Press, 1982): 287–368.

134 **an alternative hypothesis** M. M. Smith and M. I. Coates, "Evolutionary Origins of the Vertebrate Dentition: Phylogenetic Patterns and Developmental Evolution," *European Journal of Oral Sciences* 106, suppl. 1 (January 1998): 482–500, https://doi.org/10.1111/j.1600-0722.1998.tb02212.x.

135 **a PhD dissertation project** W. A. Schutt Jr., "Digital Morphology in the Chiroptera: The Passive Digital Lock," *Acta Anatomica* 148, no. 4 (1993), 219–27, https://doi.org/10.1159/000147544.

136 **in what is now the Czech Republic** V. Vaškaninová et al., "Marginal Dentition and Multiple Dermal Jawbones as the Ancestral Condition of Jawed Vertebrates," *Science* 369 (July 10, 2020): 211–16, https://www.science.org/doi/10.1126/science.aaz9431.

136 **an "inside and out" model** G. J. Fraser et al., "The Odontode Explosion: The Origin of Tooth-Like Structures in Vertebrates," *BioEssays* 32, no. 9 (September 2010): 808–17, https://www.ncbi.nlm.nih.gov/pmc/articles/PMC3034446.

Chapter 10: A Painless Guide to Tooth Basics

144 **after the oral jaws secure the prey** National Science Foundation, "Moray Eels Are Uniquely Equipped to Pack Big Prey into Their Narrow Bodies," news release 07-113, https://www.nsf.gov/news/news_images.jsp?cntn_id=109985&org=NSF.

Chapter 11: Of Fish and Frogs

151 **frogs had a much higher rate of edentulism** D. J. Paluh et al., "Rampant Tooth Loss Across 200 Million Years of Frog Evolution," *eLife*, June 1, 2021, https://doi.org/10.7554/eLife.66926.

151 **"have lost their teeth more times"** Sofia Andrade, "What Is Going On with Frog Teeth?" *Slate*, June 1, 2021, https://slate.com/technology/2021/06/frog-teeth-evolution-loss-prey.html.

152 **134 out of 429 frog species** Paluh et al., 2.

152 **"in good health and well developed"** A. Mulas et al., "Living Naked: First Case of Lack of Skin-Related Structures in an Elasmobranch, the Blackmouth Catshark (*Galeus melastomus*)," *Journal of Fish Biology* 97, no. 4 (October 2020): 1252–56, https://doi.org/10.1111/jfb.14468.

154 **more than twenty separate occasions** Paluh et al., e66926.

154 **not one of the more than 350 extant species** T. Davit-Béal, A. S. Tucker, and J. Sire, "Loss of Teeth and Enamel in Tetrapods: Fossil Record, Genetic Data and Morphological Adaptations," *Journal of Anatomy* 214, no. 4 (April 2009): 483, https://doi.org/10.1111/j.1469-7580.2009.01060.x.

Chapter 12: Dinosaurs, Turtles, Birds, and Dresser Drawers

163 **teeth in a sixteen-day-old chick embryo** "Mutant Chickens Grow Teeth," *Science*, February 21, 2006, https://www.science.org/content/article/mutant-chickens-grow-teeth.

163 **similar experiments by others** Reviewed by Davit-Béal et al., 480.

164 **That structure was a beak** Davit-Béal et al., 482.

164 **the first beaks covered only the tip of the jaw** S. Wang et al., "Heterochronic Truncation of Odontogenesis in Theropod Dinosaurs Provides Insight into the Macroevolution of Avian Beaks," *Proceedings of the National Academy of Sciences* (September 2017): 1–6, https://doi.org/10.1073/pnas.1708023114.

164 **the ancient turtle *Eorhynchochelys sinensis*** C. Li et al., "A Triassic Stem Turtle with an Edentulous Beak," *Nature* 560 (2018): 476–79, https://doi.org/10.1038/s41586-018-0419-1.

164 **"a half-beak, half-toothed jaw"** "Turtles Weren't Always Toothless: 'Missing Link" in Turtle Evolution Found," Agence France-Presse, August 23, 2018, https://www.firstpost.com/tech/science/turtles-werent-always-toothless-missing-link-in-turtle-evolution-found-5025101.html.

165 **beak size of the medium ground finch** Jonathan Weiner, *The Beak of the Finch* (New York: Vintage, 1995), 332.

167 **giant herbivores known as sauropods** J. O. Calvo, "Gastroliths in Sauropod Dinosaurs," *GAIA—Ecological Perspectives on Science and Society* 10 (December 1994): 205–08.

167 resembled huge fermentation vats O. Wings and P. M. Sander, "No Gastric Mill in Sauropod Dinosaurs: New Evidence from Analysis of Gastrolith Mass and Function in Ostriches," *Proceedings of the Royal Society B* 274 (2007): 635–40, https://doi.org/10.1098/rspb.2006.3763.

168 "birds have come down to us through the Dinosaurs" O. C. Marsh, "Introduction and Succession of Vertebrate Life in America," *American Journal of Science*, 3rd series, 14 (1877): 352.

168 "Surely there is nothing very wild or illegitimate" T. H. Huxley, "On the Animals Which Are Most Nearly Intermediate Between Birds and Reptiles," *Annals and Magazine of Natural History*, 4th series, 2 (1888): 74.

169 John Ostrom J. Ostrom, "The Ancestry of Birds," *Nature* 242, no. 5393 (1972): 136, https://doi.org/10.1038/242136a0.

169 Mark Norell A. H. Turner et al., "A Basal Dromaeosaurid and Size Evolution Preceding Avian Flight," *Science* 317, no. 5843 (September 7, 2007): 1378–81, https://www.science.org/doi/10.1126/science.1144066.

Chapter 13: Toothless Mammals, from Anteaters to Whales

172 they lay eggs H. L. Green, "A Description of the Egg Tooth of *Ornithorhynchus*, Together with Some Notes on the Development of the Palatine Processes of the Premaxillae," *Journal of Anatomy* 64, part 4 (July 1930); 512–22.1.

175 pangolins have a gizzard-like region of their stomachs W. J. Krause and C. R. Leeson, "The Stomach of the Pangolin (*Manis pentadactyla*) with Emphasis on the Pyloric Teeth," *Acta Anatomica* 88 (1974): 1–10, https://doi.org/10.1159/000144218.

176 "In the ancestors of mysticetes" Davit-Béal et al., 492.

177 "A man got those two things he's got it all" James Brown, *James Brown: The Godfather of Soul* (New York: Macmillan, 1986).

Chapter 14: A Man of Few Words . . . and Fewer Teeth

182 only 12 of them showed evidence of dental work D. K. Whittaker, "Dental Aspects of the Spitalfields Exhumations," *Dental Historian* 21 (1991): 30–43; D. K. Whittaker and A. S. Hargreaves, "Dental Restorations and Artificial Teeth in a Georgian Population," *Dental History* 7 (1991): 371–76, https://doi.org/10.1038/sj.bdj.4807727.

182 Paul Revere (who occasionally dabbled in dentistry) M. F. Nola, "Paul Revere and Forensic Dentistry," *Military Medicine* 181, no. 7 (July 2016): 714–15, https://doi.org/10.7205/MILMED-D-16-00044; "The Real Story of Paul Revere's Ride," Paul Revere House, https://www.paulreverehouse.org/the-real-story.

183 Revere and several of Warren's relatives Richard Barnett, *The Smile Stealers: The Fine and Foul Art of Dentistry* (London: Thames & Hudson, 2017), 202; Jacqueline Gase, "Paul Revere and Joseph Warren: An Early Case of Forensic Identification," *Micrograph*, National Museum of Health and Medicine, https://medicalmuseum.health.mil/micrograph/index.cfm/posts/2019/paul_revere_and_joseph_warren.

183 All four sets of George Washington's surviving dentures J. Van Horn, "George Washington's Dentures: Disability, Deception, and the Republican Body," *Early American Studies* 14, no. 1 (Winter 2016), 2–47, https://doi.org/10.1353/eam.2016.0000.

184 Adolf Hitler's antemortem and postmortem dental status R. F. Sognnaes and F. Ström, "The Odontological Identification of Adolf Hitler: Definitive Documentation by X-Rays, Interrogations and Autopsy Findings," *Acta Odontologica Scandinavica* 31, no. 1 (February 1973): 43–69, https://doi.org/10.3109/00016357309004612.

184 precise models of the dental relics R. F. Sognnaes, "George Washington's Bite," *Journal of the California Dental Association* 4, no. 6 (1976): 36–40.

184 "gold springs, gold pins" Sognnaes, "George Washington's Bite," 40.

184 "America's first native-born dentist" Van Horn, 5.

185 "from Washington's own mouth" Sognnaes, "George Washington's Bite," 36.

185 more like performance artists E. Fee et al., "The Tooth Puller [*L'arracheur de dents*]," *American Journal of Public Health* 92, no. 1 (January 2002): 35, https://doi.org/10.2105/AJPH.92.1.35.

186 preserving this treasure in a tiny glass case "The Trouble with Teeth," George Washington's Mount Vernon, https://www.mountvernon.org/george-washington/health/washingtons-teeth.

187 "very pern[i]cious to the teeth" W. W. Abbot and Edward G. Lengel, eds., *The Papers of George Washington: Retirement Series*, vol. 3, September 1798–April 1799 (Charlottesville: University of Virginia Press, 1999), 289–91.

187 the new concept of daily tooth care T. G. H. Drake, "Antiques of Medical Interest: Tooth Brush Set, Silver, London, 1799," *Journal of the History of Medicine and Allied Sciences* 2, no. 1 (1947): 48–50, https://doi.org/10.1093/jhmas/II.1.48.

187 urine as mouthwash N. Powers, "Archaeological Evidence for Dental Innovation: An Eighteenth Century Porcelain Dental Prosthesis Belonging to Archbishop Arthur Richard Dillon," *British Dental Journal* 201, no. 7 (October 7, 2006), 459–63, https://doi.org/10.1038/sj.bdj.4814117.

188 the French dentist Jean-Pierre Le Mayeur M. V. Thompson. "And Procure for Themselves a Few Amenities: The Private Life of George Washington's Slaves," *Virginia Cavalcade* 48, no. 4 (Autumn 1999): 183, https://archive.org/details/sim_virginia-cavalcade_autumn-1999_48_4/page/n3/mode/2up.

188 "In a drawer, in the Locker of the Desk" "From George Washington to Lund Washington, 25 December 1782," *Founders Online*, National Archives, https://founders.archives.gov/documents/Washington/99-01-02-10299.

189 "It was a bloody smile" Victor Hugo, *Les Misérables*, trans. Isabel F. Hapgood (New York: Thomas Y. Crowell, 1887), book 5, chap. X, https://www.gutenberg.org/files/135/135-h/135-h.htm#link2HCH0049.

189 the battlefields of the American Civil War Marshall J. Becker and Jean MacIntosh Turfa, *The Etruscans and the History of Dentistry: The Golden Smile Through the Ages* (London: Routledge, 2017), 155.

190 "and not one succeeded" Charles R. E. Koch, ed., *Koch's History of Dental Surgery* (Fort Wayne, IN, 1910), "Lemaire" 3: 5–8. Referenced by Weinberger, *An Introduction to the History of Dentistry in America*, 161.

190 contracting syphilis as well J. J. Ross, "Reply to Gladstein" [letter], *Clinical Infectious Diseases* 41 (2005):128–29.

190 first successful artificial insemination Manasseh Ngigi, "The Natural History of Human Teeth—John Hunter," UT Health San Antonio Libraries, https://library.uthscsa.edu/2015/03/the-natural-history-of-human-teeth-john-hunter.

191 "*not that I believe it possible to transplant an infection*" John Hunter, *A Practical Treatise on the Diseases of the Teeth: Intended as a Supplement to the Natural History of Those Parts* (London: J. Johnson, 1778), 100.

191 "I succeeded but once" Hunter, *A Practical Treatise*, 112.

192 Everard Home burned them K. A. Kapp and G. E. Talboy, "John Hunter, the Father of Scientific Surgery," American College of Surgeons Poster Competition (2017): 34–41.

193 "Transplanted teeth can never recover life" M. Blackwell, "'Extraneous Bodies': The Contagion of Live-Tooth Transplantation in Late-Eighteenth-Century England," *Eighteenth Century Life* 28, no. 1 (2004): 61, https://doi.org/10.1215/00982601-28-1-21.

194 "when it is baked has every desirable advantage" J. Cigrand, *The Rise, Fall and Revival of Dental Prosthesis* (Chicago: Periodical Publishing, 1893), 139.

194 a dining table he had also invented Blackwell, 55.

194 underwent tooth transplantation procedures Michael Coard, "George Washington's Teeth Not from Wood but Slaves," *Philadelphia Tribune*, February 24, 2018, https://www.phillytrib.com/commentary/coard-george-washingtons-teeth-not-from-wood-but-slaves/article_f9f31911-bcdc-53ac-b86d-26aba781824d.html.

195 "tobacco, tea, coffee, port wine, etc." Sognnaes, "George Washington's Bite," 36.

196 a peculiar notation made in May 1784 Thompson, 178–90.

196 **"Cash p[ai]d on Acc[oun]t of Genrl. Washington"** "George Washington and Teeth from Enslaved People," Mount Vernon, https://www.mountvernon.org/george-washington/health/washingtons-teeth/george-washington-and-slave-teeth.

196 **amount paid for the nine teeth** Thompson, 183.

198 **the price of "white gold"** Barnett, 44.

200 **mercury poisoning has been implicated** Laura Lane, "Amalgam Wars: Part I," DrBicuspid.com, https://www.drbicuspid.com/home/article/15355697/amalgam-wars-part-i.

200 **"every tooth in my head became so loose"** "George Washington's Troublesome Teeth," *Lives & Legacies* (blog), May 18, 2017, https://livesandlegaciesblog.org/2017/05/18/george-washingtons-troublesome-teeth.

202 **"selling their teeth to dentists"** Thompson, 183.

203 **almost certainly constructed by a different dentist** Weinberger, 164.

204 **approximately three hundred of them** Thompson, 179–80.

204 **"a set of false teeth inserted"** "Face: The Portrait," George Washington: A National Treasure, Smithsonian Portrait Gallery, https://georgewashington.si.edu/portrait/face.html.

206 **8.8 percent of young American military enlistees** "Dental Standards for Military Service," AMEDD Center of History & Heritage, U. S. Army Medical Center of Excellence, https://achh.army.mil/history/corps-dental-wwii-chaptervi-wwii.

206 **a roll call of Hollywood types** Brennan Kilbane, "The History of the Big, Bright American Smile," *Allure*, December 15, 2020, https://www.allure.com/story/american-smile-good-teeth; Mackenzie Dawson, "How Judy Garland Launched America's Quest for the Perfect Smile," *New York Post*, March 18, 2017, https://nypost.com/2017/03/18/how-judy-garland-launched-americas-quest-for-the-perfect-smile.

207 **the bright-white American smile** Mary Otto, *Teeth: The Story of Beauty, Inequality, and the Struggle for Oral Health in America* (New York: New Press, 2017).

207 **the involuntary nature of the toil and hardship** Aaron O'Neill, "Black and Slave Population of the United States from 1790 to 1880," Statista, June 21, 2022, https://www.statista.com/statistics/1010169/black-and-slave-population-us-1790-1880.

207 **"his Teeth stragling and fil'd sharp"** "Advertisement for Runaway Slaves, 11 August 1761," *Founders Online*, National Archives, https://founders.archives.gov/documents/Washington/02-07-02-0038. [Original source: *The Papers of George Washington*, Colonial Series, vol. 7, *1 January 1761–15 June 1767*, eds. W. W. Abbot and Dorothy Twohig (Charlottesville: University of Virginia Press, 1990), 65–68.].

207 **corn consumption** P. M. Lambert, "Infectious Disease Among Enslaved African Americans at Eaton's Estate, Warren County, North Carolina, ca. 1830–1850," *Memórias do Instituto Oswaldo Cruz* 101, suppl. 2 (December 2006), https://doi.org/10.1590/S0074-02762006001000017.

208 tiny transverse grooves called perikymata I. Roa and M. del Sol, "Perikymata: A Non-Existent Term. A Scientific Literature Invention? Terminology Analysis and Proposal," *International Journal of Morphology* 35, no. 4 (2017): 1230–32, http://dx.doi.org/10.4067/S0717-95022017000401230.

209 Hypoplasias have been recorded Tanya M. Smith and Christina Warinner, "Developmental, Evolutionary and Behavioural Perspectives on Oral Health," *Palaeopathology and Evolutionary Medicine*, Kimberly A. Plomp et al., eds. (Oxford: Oxford University Press, 2022), 96–97.

209 "the poorest populations ever studied" Richard H. Steckel, "The African American Population of the United States, 1790–1920," Michael R. Haines and Richard H. Steckel, eds., *A Population History of North America* (Cambridge: Cambridge University Press, 2000), 449.

209 died at rates of 350 and 200 per thousand R. H. Steckel, "A Dreadful Childhood: The Excess Mortality of American Slaves," *Social Science History* 10, no. 4 (Winter 1986): 427–65, https://doi.org/10.2307/1171026.

209 mulberry molars . . . and Hutchinson's teeth Lambert, 114.

210 hypercementosis Lambert, 114.

Chapter 15: Jaw Jewelry, Pliers, and Pelicans

212 *The Etruscans and the History of Dentistry* Becker and Turfa, *The Etruscans and the History of Dentistry*, 415.

213 to advertise the wealth and lofty societal status X. Riaud, "History of Dental Implantology," *Austin Journal of Dentistry* 4, no. 4 (2017): 1080.

213 Etruscans had "rather good dental health" Becker and Turfa, 100.

214 had their healthy teeth yanked Marshall J. Becker, "Etruscan Female Tooth Evulsion: Gold Dental Appliances as Ornaments," *Practices, Practitioners and Patients: New Approaches to Medical Archaeology*, Patricia Anne Baker and Gillian Carr, eds. (Oxford: Oxbow Books, 2017), 235–57.

215 inlays made of finely crafted stones Becker and Turfa, 111–12; Marshall Becker (pers. comm.).

216 in the drain of a *taberna* Becker and Turfa, 15, 19–20.

216 Emperor Claudius's physician Becker, 38–41.

216 "The first known illustrations of instruments" Becker and Turfa, 56.

216 a painting by Guido Reni Guido Reni, *The Martyrdom of Saint Apollonia*, 1600–1603, oil on copperplate, Museo del Prado, Madrid, Spain, https://www.museodelprado.es/en/the-collection/art-work/the-martyrdom-of-saint-apollonia/2042a139-8eb6-41c2-99ef-bdc6757c8f6c.

217 patron saint of dentists Barnett, 35.

217 pelicans were used by coopers Barnett, 48.

219 died in London from dental issues Weinberger, 12.

220 **written during the Tang dynasty** A. Czarnetzki and S. Ehrhardt, "Re-Dating the Chinese Amalgam-Filling of Teeth in Europe," *International Journal of Anthropology* 5, no. 4 (1990): 325–32.

220 **the European blend** Becker and Turfa, 147; Czarnetzki and Ehrhardt, 326

221 **denouncing the "use of amalgams"** American Society of Dental Surgeons, *American Journal of Dental Science* (Cambridge, MA: Harvard University, 1845): 169–70.

222 **"mercury serves no healthful purpose"** M. E. Mortensen, "Mysticism and Science: The Amalgam Wars, Journal of Toxicology," *Clinical Toxicology* 29, no. 2 (1991): vii–xii.

223 **an extremely high concentration of fluoride** Peter Meiers, "The Bauxite Story: A Look at ALCOA," Fluoride-History.de, http://www.fluoride-history.de/bauxite.htm.

224 **the removal of fluoride from drinking water** C. Till and R. Green, "Controversy: The Evolving Science of Fluoride: When New Evidence Doesn't Conform with Existing Beliefs," Pediatric Research 90, no. 5 (November 2021):1093–95, https://doi.org/10.1038/s41390-020-0973-8.

224 **IQ scores reduced by three to five points** R. Green et al., "Association Between Maternal Fluoride Exposure During Pregnancy and IQ Scores in Offspring in Canada," *JAMA Pediatrics* 173 (2019): 940–48, https://doi.org/10.1001/jamapediatrics.2019.1729.

224 **health concerns over fluoridated tap water** "Community Water Fluoridation," Centers for Disease Control and Prevention, January 15, 2020, https://www.cdc.gov/fluoridation.

Chapter 16: Tooth Worms

226 **medical and dental folklorist B. R. Townend** B. R. Townend, "The Story of the Tooth-Worm," *Bulletin of the History of Medicine* 15, no. 1 (January 1944): 37–58, http://www.jstor.org/stable/44442797.

226 **more recent papers on the topic** W. E. Gerabek, "The Tooth-Worm: Historical Aspects of a Popular Medical Belief," *Clinical Oral Investigations* 3, no. 1 (March 1999): 1–6, https://doi.org/10.1007/s007840050070.

227 **"the worm theory" of dental disease** Townend, 42.

227 **nineteenth-century Cherokee rituals** Townend, 43.

227 **"from the eggs of flies and other insects"** Townend, 44.

227 **anatomist and Egyptologist Grafton Smith** B. R. Townend: 40.

229 **smoke out tooth worms** M. Al Hamdani and M. Wenzel, "The Worm in the Tooth," *Folklore* 77, no. 1 (Spring 1966): 60–64, http://www.jstor.org/stable/1258921.

230 **In a related treatment** Al Hamdani and Wenzel, 61.

231 "Roast a piece of garlic" Barnett, 33.

231 The treatments he unearthed Townend, 55.

232 early-twentieth-century German custom Gerabek, 5.

232 the 1728 publication of his magnum opus Pierre Fauchard, *Le Chirurgien Dentiste, ou Traité des Dents* (Paris: Pierre-Jean Mariette, 1728).

237 possible causative agent in the development of Alzheimer's disease P. I. Diaz, P. S. Zilm, and A. H. Rogers, "*Fusobacterium nucleatum* Supports the Growth of *Porphyromonas gingivalis* in Oxygenated and Carbon-Dioxide -Depleted Environments," *Microbiology* 148 (2002): 467–72, https://doi.org/10.1099/00221287-148-2-467.

Chapter 17: Wisdom Teeth, Baby Teeth, and the Tooth Fairy

241 "liberated from lengthy bouts of daily mastication" A. V. Casteren et al., "The Cost of Chewing: The Energetics and Evolutionary Significance of Mastication in Humans," *Sciences Advances* 8, no. 33 (August 17, 2022), https://doi.org/10.1126/sciadv.abn8351.

242 lack an extra set of molars Tanya M. Smith, *The Tales Teeth Tell* (Cambridge, MA: MIT Press, 2018), 60.

242 a relatively recent evolutionary response Smith, 61–63.

243 "may be compared with the letters in a word" Charles Darwin, *The Origin of Species by Means of Natural Selection, or the Preservation of Favoured Races in the Struggle for Life* (London: John Murray, 1872), 350.

248 "Many a refractory child" Lillian Brown, "Tooth Fairy," Practical Housekeeper's Own Page, *Chicago Tribune*, September 27, 1908, https://www.newspapers.com/newspage/354963876.

248 Evidence of Nellie May's answer "My Visit to Fairyland," A Page for Children by Children, *Courier-Journal*, May 14, 1911, https://www.newspapers.com/image/118824932.

248 the average value of a lost tooth Delta Dental, "Delta Dental Poll Finds Tooth Fairy Welcomed into Most U.S. Homes," news release, August 17, 2023, https://www.deltadental.com/us/en/tooth-fairy/press-release.html.

Chapter 18: To Infinity and Beyond . . . the Dental Chair

251 "synthetic biomaterials" Anthony Cuthbertson, "Dental Fillings Heal Teeth with Stem Cells, *Newsweek*, July 4, 2016, https://www.newsweek.com/dental-fillings-heal-teeth-stem-cells-harvard-cavities-477415.

252 levels of *P. gingivalis* were significantly higher S. S. Dominy et al., "*Porphyromonas gingivalis* in Alzheimer's Disease Brains: Evidence for Disease Causation and Treatment with Small-Molecule Inhibitors," *Science Advances* 5, no. 1 (January 23, 2019), https://www.science.org/doi/10.1126/sciadv.aau3333.

252 slowed cognitive decline in a subset of study participants GlobalData Healthcare, "Despite New Biomarker Data for AD Drug COR388, Clinical Efficacy Remains Unproven," editorial, Clinical Trials Area, March 24, 2022, https://www.clinicaltrialsarena.com/comment/biomarker-data-cor388-clinical-efficacy.

254 a transition to a broader diet Julie Russ, "Ancient Human Ancestor's Teeth Reveal Diverse Diet," June 3, 2013, ASU News, Arizona State University, https://news.asu.edu/content/ancient-human-ancestors-teeth-reveal-diverse-diet.

254 Denisova Cave in Siberia D. Reich et al., "Genetic History of an Archaic Hominin Group from Denisova Cave in Siberia," *Nature* 468 (2010): 1053–60, https://doi.org/10.1038/nature09710.

254 neither Neanderthal nor modern human V. Slon et al., "The Genome of the Offspring of a Neanderthal Mother and a Denisovan Father," *Nature* 561 (2018): 113–16.

255 the best match for the tooth F. Demeter et al., "A Middle Pleistocene Denisovan Molar from the Annamite Chain of Northern Laos," *Nature Communications* 13 (2022): 2557.

255 this bit of physical evidence Michael Price, "Ancient DNA Puts a Face on the Mysterious Denisovans, Extinct Cousins of Neanderthals," September 19, 2019, *Science*, https://www.science.org/content/article/ancient-dna-puts-face-mysterious-denisovans-extinct-cousins-neanderthals.

257 the list of potential competitive advantages K. Harvati, "What Happened to the Neanderthals?" Nature Education Knowledge 3, no. 10 (2012):13, https://www.nature.com/scitable/knowledge/library/what-happened-to-the-neanderthals-68245020.

257 dental calculus scraped from the teeth of Neanderthal skeletons A.G. Henry, A.S. Brooks, and D. R. Piperno, "Microfossils in Calculus Demonstrate Consumption of Plants and Cooked Foods in Neanderthal Diets (Shanidar III, Iraq; Spy I and II, Belgium)," *Proceedings of the National Academy of Sciences* 108, no. 2 (December 27, 2010): 486–91, https://doi.org/10.1073/pnas.1016868108.

257 information on diet T. Juhola et al., "Phytoliths, Parasites, Fibers, and Feathers from Dental Calculus and Sediment from Iron Age Luistari Cemetery, Finland," *Quaternary Science Reviews* 222 (2019): 105888, https://doi.org/10.1016/j.quascirev.2019.105888.

258 this population of Iron Age Finlanders Anna Goldfield "Do I Have Microremains in My Teeth?" Field Trips, *Sapiens*, May 20, 2020, https://www.sapiens.org/biology/microremains-teeth.

259 "it's not possible to simply look at the teeth" Smith, *The Tales Teeth Tell*, 147.

259 believed that they were fish teeth C. H. Pander, *Monographie der Fossilen Fische des Silurischen Systems der Russisch-Baltischen Gouvernements* (Monograph of Fossil Fish from the Silurian Stratum of the Baltic Regions), St. Petersburg (1856).

260 a remarkably long stretch of time Yong Yi Zhen, "What Are Conodonts?" Australian Museum, July 16, 2020, https://australian.museum/learn/australia-over-time/fossils/what-are-conodonts.

261 tree-ring-like incremental growth lines Smith, *The Tales Teeth Tell.*

261 use information obtained from teeth on early life adversity K. A. Davis et al., "Teeth as Potential New Tools to Measure Early-Life Adversity and Subsequent Mental Health Risk: An Interdisciplinary Review and Conceptual Model," *Biological Psychiatry* 87, no. 6 (March 15, 2020): 502–13, https://doi.org/10.1016/j.biopsych.2019.09.030.

262 questions, and indeed hurdles, remain Davis et al. (2020): table 2, 23.

262 tooth replacement seen in alligators P. Wu et al., "Specialized Stem Cell Niche Enables Repetitive Renewal of Alligator Teeth," *Proceedings of the National Academy of Sciences* 110, no. 22 (May 28, 2013): E2009–18, https://doi.org/10.1073/pnas.1213202110.

263 tissue from embryos, hatchlings, and juvenile gators Rachel Nuwer, "Solving an Alligator Mystery May Help Humans Regrow Lost Teeth," *Smithsonian*, May 13, 2013, https://www.smithsonianmag.com/science-nature/solving-an-alligator-mystery-may-help-humans-regrow-lost-teeth-60257334.

263 Sayaka Mishima and his colleagues at Kyoto University S. Mishima et al., "Local Application of *Usag-1* siRNA Can Promote Tooth Regeneration in *Runx2*-Deficient Mice," *Scientific Reports* 11, no. 13674 (2021), https://doi.org/10.1038/s41598-021-93256-y.

263 "rescued tooth development" Tim Newcomb, "Humans Have a Third Set of Teeth. New Medicine May Help Them Grow," *Popular Mechanics*, September 4, 2023, https://www.popularmechanics.com/science/health/a44786433/humans-have-third-set-teeth. .

Epilogue: How Science Works

270 "very likely used its shovel-shaped jaw to scoop up succulent water plants" H. F. Osborn and W. Granger, "*Platybelodon grangeri*, Three Growth Stages, and a New Serridentine from Mongolia," *American Museum Novitates* 537 (June 9, 1932): 13, http://hdl.handle.net/2246/2995.

270 the creature was often depicted in books Patricia Vickers Rich et al., *The Fossil Book: A Record of Prehistoric Life* (Mineola, NY: Dover, 1996), 586; Robert J. G. Savage and M.R. Long. *Mammal Evolution: An Illustrated Guide* (New York: Facts on File, 1986), 153.

270 even museum displays "Platybelodon Grangeri of Miocene at Inner Mongolia Museum, in Hohhot, China," Wikipedia, https://en.wikipedia.org/wiki/Platybelodon#/media/File:Hohhot.inner_mongolia_museum.Platybelodon_grangeri.2.jpg.

270 paleontologist David Lambert proposed something different W. D. Lambert, "The Feeding Habits of the Shovel-Tusked Gomphotheres: Evidence from Tusk Wear Patterns," *Paleobiology* 18, no. 2 (Spring 1992): 132–47, http://www.jstor.org/stable/2400995.

271 assessed the diet of *Platybelodon* G. M. Semprebon et al., "An Examination of the Dietary Habits of *Platybelodon grangeri* from the Linxia Basin of China: Evidence from Dental Microwear of Molar Teeth and Tusks," *Palaeogeography, Palaeoclimatology, Palaeoecology* 457 (September 1, 2016): 109–16, https://doi.org/10.1016/j.palaeo.2016.06.012.

271 "a snapshot of dietary behavior" Semprebon, 111.

Acknowledgments

I WOULD LIKE to thank my agent Gillian MacKenzie for her hard work, great advice, perseverance, and patience.

I offer my sincere thanks to Amy Gash, my incredibly talented editor at Algonquin Books, for her significant and outstanding editorial input and advice, and I can't wait to see what we'll be working on next. My thanks and gratitude to publicity assistant Katrina Tiktinsky, managing editor Brunson Hoole, designer Steve Godwin, and my extremely talented copyeditor, Elizabeth Johnson, for all their hard work; and to publicity wiz Johanna Ramos-Boyer for helping secure so many wonderful media gigs for me over the years. Thanks also to Ashley Cave Himes and Jayme Boucher at Hachette Speakers Bureau, and to the entire production and marketing teams at Algonquin Books. It is an absolute pleasure working with you!

I was very lucky to have interviewed or received assistance from a long list of experts who were extremely generous with their time and their tooth-related knowledge. Much thanks and gratitude go out to Per Ahlberg, Marshall Becker, John Bertram, Gerry Carter, Nick Casewell, Vincent Dupret, Mark Engstrom, Greg Erickson, Theresa Felton, Brock Fenton, Lázaro Guevara, John Hermanson, Burton Lim, Nathan Lujan, Ross MacPhee, Jacqueline Miller, Daniel Paluh, Karen Reiss, Susan Schoelwer (who was an integral contributor to the George Washington story), William R. Schutt, Kevin Sheren, Nancy Simmons (it's good to know the Queen), Tanya Smith, Daniel Snitting, Stephen Spotte, Ian Tattersall, and Valéria Vaškaninová.

I owe a huge debt of gratitude to my friends and colleagues in the bat research community and at the American Museum of Natural History. They include Ricky Adams, Frank Bonaccorso, Emily Davis, Betsy Dumont, Neil Duncan, Julie Faure-Lacroix, Ellen Futter, Roy Horst (RIP), Kristin Jonasson, Mary Knight, Tom Kunz (RIP), Gary Kwiecinski, Shahroukh Mistry, Farouk Muradali, Mark Norell, Mike Novacek, Scott Pedersen, Bruce Patterson, Nereida Rodriguez, Elizabeth Taylor, and Rob Voss.

I've been fortunate to have had several incredible mentors, none more important than John Hermanson. Among many other things, John taught me to think like a scientist, as well as the value of figuring stuff out for myself.

A very special thanks to my great friend, collaborator, confidant, and coconspirator Leslie Nesbitt Sittlow.

As usual, my dear friends Darrin Lunde and Patricia J. Wynne were instrumental in helping me develop this project from a vague idea into a finished book. A million thanks also to Patricia for all the amazing figures she drew (not to mention the spot-on advice). As always, I can't wait for our next project together.

I'd also like to express my deep gratitude to Cornell University zoologist and teacher extraordinaire Howard Evans (1922–2023) for showing me that it's perfectly okay to be enthusiastic about vampire bats, blue whale hearts, and mating sturgeons mistaken for the Loch Ness Monster.

A special thank you goes out to my teachers, readers, and supporters at the Southampton College Summer Writers Conference, especially Bob Reeves, Bharati Mukherjee (RIP), and Clark Blaise.

At Southampton College (RIP) and LIU Post, thanks and gratitude go out to Kim Cline, Art Goldberg (RIP), Alan Hecht, Kent Hatch, Mary Lai (RIP), Karin Melkonian, Kathy Mendola, Glynis Pereyra, Howard

Reisman, and Steve Tettelbach. Thanks also to my LIU Post graduate students and teaching assistants.

Sincere thanks go out to Bob Adamo (RIP), John Bodnar, Chris Chapin, Kitty Charde, Kristi Ashley Collom, Alice Cooper, Justin Esposito, Suzanne Finnamore Luckenbach (who predicted it all), John Glusman, Chris Grant, Kathy and Brian Kennedy, Seb Kvist, Christian Lennon and Erin Nicosia-Lennon, Bob Lorzing, the legendary and wonderful literary agent Elaine Markson (RIP), Maceo Mitchell, Carrie and Dan McKenna, Val Montoya, Ellen Morris-Knower, the Pedersen family and various offshoots, Ashley, Kelly, and Kyle Pellegrino, Don Peterson, Adam Phillips, Doranne Phillips Telberg, Gerard, Oda, and Dominique Ramsawak, Miranda Rice, Carole Roble, Jerry Ruotolo, Kirsten Sanford, Brandy Schillace (*Peculiar Book Club*), Laura Schlecker, Lynn Swisher, Frank Trezza, Carol Trezza (RIP), Katherine Turman (*Nights with Alice Cooper*), Dorothy Wachter (RIP), Mindy Weisberger, "Pirate" Mike Whitney, and Jennifer Wilkinson.

Finally, eternal thanks and love go out to my wonderful wife, Janet Schutt, and my son, (Dr. William R. Schutt, a.k.a. Billy Schutt), for their patience, love, encouragement, and unwavering support. Thanks also to my grandparents (Angelo and Millie DiDonato), aunts and uncles (including my Aunt Ei and all my Aunt Roses), cousins, nieces, and nephews, and, most of all, my parents, Bill and Marie G. Schutt.

Index